CoalDust
grins

PORTRAITS OF CANADIAN COAL MINERS

CoalDust grins

PORTRAITS OF CANADIAN COAL MINERS

PHOTOGRAPHY AND TEXT

BY

LAWRENCE CHRISMAS

EDITED BY KERRY McARTHUR

Cambria Publishing

CANADIAN CATALOGUING IN PUBLICATION DATA

Chrismas, Lawrence.
CoalDust Grins
Portraits of Canadian Coal Miners

Includes index.
ISBN 0-9697023-1-0

1. Coal miners – Canada – Portraits.   2. Coal miners – Canada – Biography.
3. Coal mines and mining – Canada.   I. Title.
TR681.M64C57 1998 622'.334' 092271   C98-910916-X

Printed and bound in Canada

Published by

CAMBRIA PUBLISHING

Box 61083 Kensington PO
Calgary, Alberta T2N 4S6

E-mail: publisher@cambriapublishing.com
Web site: www.cambriapublishing.com

CONTENTS

INTRODUCTION BY LAWRENCE CHRISMAS

It's not that I'm an outsider in awe of coal miners – I am a legitimate member of the coal industry. I've not actually dug coal but many times my "coaldust face" has been unrecognizable after a trip underground. Through a professional career and my personal involvement in photographing Canadian coal miners, I have visited surface and underground mines in Canada, the United States, England, Wales, Belgium, France, Germany and South Korea.

Fresh out of university 30 years ago, I started with the Department of Energy, Mines and Resources, working as a coal resource specialist. The boss said, "Visit all the coal mines in the Maritimes and learn about coal mining." Off I went for a six-week tour of all operating coal mines in New Brunswick and Nova Scotia.

That was in 1968. My first trip underground was in a little family-operated mine in Minto, New Brunswick – what an introduction. The seam was about 18 inches thick; the roadway was only three feet high. From the shaft bottom, I was on my hands and knees for 400 feet until I reached the miners. They were on their sides with pick and shovel, digging – their shoulders brushing the roof of the seam. It was an amazing and claustrophobic scene.

The pit boss guiding me through the mine quietly enjoyed my discomfort. When he stopped to let me catch up, he said, "Let's have a smoke," and he lit them from the open-flame carbide lamp on his hardhat. I thought I'd never get out of that little coal mine alive.

Nonetheless, the real charisma of coal mining began for me on that same day. By the time I reached surface, I was ready and anxious to go underground again. Over the next few weeks, I entered both the Princess and No. 26 Collieries in Cape Breton Island, Nova Scotia. Both these mines extended up to five miles beneath the ocean floor and I spent much of my time trying not to think about the Atlantic Ocean over my head and how far I was from the mine entrance.

I remember climbing around the roof supports and watching the men operating the shearers (large machines for cutting coal at rapid speeds.) The dust was unbelievable and, in those days, there were no masks to filter the coal dust. At the end of the day in the washhouse, I spent a lot of time in the shower trying to wash off the dust. A burly miner beside me said, "Hey, buddy, you want your back washed?" I think I jumped 10 feet! But I learned later that the miners often washed each others' backs.

During that same trip, I went into the McBean Mine in Stellarton to study a hand-loaded longwall with about 40 or 50 miners undercutting, drilling and blasting their sections. Word came that a few miners were ready to shoot so we moved back a ways. There was a muffled boom, followed by a cloud of smoke, dust and men scrambling for cover under timbers. The ground shook beneath my feet and small chunks of rock fell from the roof.

Later, I was told I'd just experienced a "bump," or minor earthquake, that had been set off by the blast. No one was hurt at the time but the encounter helped me understand the terrible bumps described by so many old-time miners who'd lived through them at Coal Creek, British Columbia and Springhill, Nova Scotia.

After I began more frequent visits to coal mines across Canada, I realized I should be photographing the remarkable people of this industry. The decision coincided with a 1979 trip to Canmore, Alberta when the mine owners suddenly closed the venerable Canmore Mines. The community of Canmore was in shock: after all, this mine had operated nearly 100 continuous years and fostered two to three generations of coal-mining families. During that week, I photographed about a dozen old-time Canmore miners, many owned Napoleon mantle clocks that distinguished them with 50 years of service to the mine.

It was the beginning of a major photo-documentary project. Twenty years later, I continue to photograph and interview Canadian coal miners.

This collection contains portraits of many underground miners who began their careers at an early age. The mines they worked are all closed; the techniques they used no longer exist. The industry has changed substantially. Today, you can count the number of Canadian underground mines on one hand. In western Canada, the trend is towards surface mining.

Surface mining hasn't been around as long as its underground counterpart, but the surface coal miner nonetheless has a distinguishable personality. His (and increasingly, her) knowledge is now of giant draglines, massive electric shovels, monstrous 150-tonne trucks and satellite communications. In other words, this is a highly-educated, skilled and productive workforce.

Over the years I have been a privaledged visitor to many of the Canadian draglines, a symbol of surface mining. What a thrill to climb the long staircase to the operators cab and observe the delicate skills of the operators and the enormous power of the draglines.

But I continue to be struck by the similarities between both underground and surface miners, from British Columbia to Nova Scotia. I think it's fair to say that the majority of men and women in coal mining today enjoy the work, comradeship, community life, high pay and job security as much as their retired counterparts. It is still a significant and important industry in the country: in 1997, Canadian coal miners produced 75 million tonnes of coal. When I began this photography collection, coal production was less than half the current amount.

When I first met John Fry (CoalDust Grins frontispiece), I'd been told he was reserved, difficult to get to know. Four hours later, I'd run out of tape for my tape recorder. Born in Wales, John came with his family to Canada when he was seven years of age. At 17, he had to sneak to work at the Hillcrest-Mohawk Mine to avoid "being skinned alive" by his mother. He was a hard-nosed miner, extremely clever and well-informed. During the early '50s, he decided coal was finished in the Crowsnest Pass, so he moved to British Columbia to work in the hardrock mines.

When the Sparwood and Coleman mines continued to show signs of life, he moved back to Blairmore and underground coal mining. He loved climbing and exploring caves in the surrounding mountains, and he had a passion for painting. I visited John many times over the years. He told me, "In coal mining, every day is different – that's what I like about it. You can't say it's boring. Let's put it this way: if you could rerun the whole damn thing like a tape, I'd run it all back the same way, the same bloody way."

This book, CoalDust Grins, has a companion music album entitled CoalDust Grins – A Musical Portrait. Produced with the guidance of Calgary music producer Tim Williams, the CD contains 12 songs by talented singer/songwriters residing in Western Canada.

The idea to produce a musical response to the book was a natural outcome of the project. In 1995, Calgary singer /songwriter Cathy Miller and I began work on a series of songs about Canadian coal miners and their lives based on my portraits and their biographies. When I photograph the miners, I conduct a recorded interview with each of them, then draft a biographical sketch to accompany the image. Some of the stories from this book were so evocative that they suggested lyrics and, by extension, songs. In fact, the title of my book – CoalDust Grins – was suggested by "The Seam," a song by Ian MacDonald and myself about the miner Archie MacDonald who eventually became the mine manager at Nova Scotia's Princess Colliery.

The whole process of collaborating with the singer/songwriters has been an enjoyable side venture to the photographic documentation of coal miners and the writing of this book.

DEDICATION

To my mother and unwavering supporter
Elana Halisky Chrismas

and to

the preservation of our
Mining History and Heritage

COAL MINERS

OF BRITISH COLUMBIA
AND YUKON

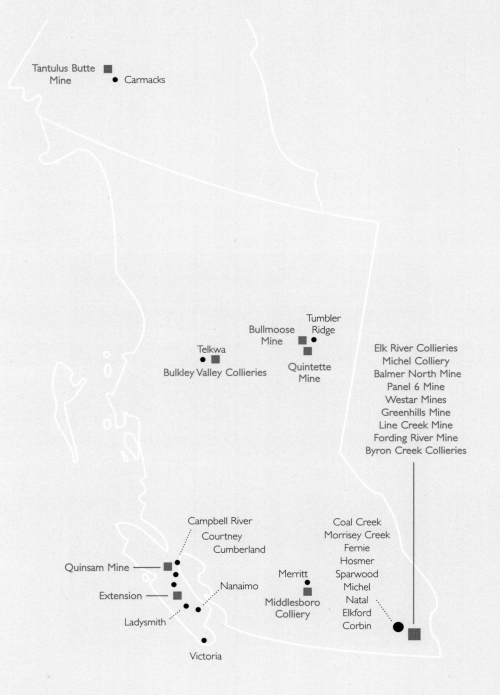

Tantulus Butte
Mine
• Carmacks

Tumbler
Ridge
Bullmoose
Mine
Telkwa •
Bulkley Valley Collieries
Quintette
Mine

Elk River Collieries
Michel Colliery
Balmer North Mine
Panel 6 Mine
Westar Mines
Greenhills Mine
Line Creek Mine
Fording River Mine
Byron Creek Collieries

Campbell River
Courtney
Cumberland
Quinsam Mine
Merritt
Coal Creek
Morrisey Creek
Fernie
Hosmer
Sparwood
Extension
Nanaimo
Michel
Middlesboro
Colliery
Natal
Elkford
Corbin
Ladysmith

Victoria

• PHOTOGRAPH LOCATION   ■ COAL MINE LOCATION

MARY AND GEORGE EDWARDS

I am 91 years old and if I live to March 15, I'll be 92. I was born and raised in Nanaimo. I lost me mother when I was two weeks old and I lost me father before I could remember him. That's how I was brought up, by the hair on my head, and I had to do it for myself.

When I got married, I never had enough money to go to Vancouver. That's how it was here. We have been married for more than 70 years.

If you don't know what hard work is, you can talk to a coal miner and he can tell you what hard work is. When you went in the mine, you could see the steam rising off the miners' backs. They were doubled up all day.

I'll tell you one little subject, me boy. Before the unions came here, it was hell. The men would get into little arguments: the coal digger thought he was right and the boss thought he was right. Well, you would go to work and the boss would say nothing to you. Or you'd go back to the lamp cabin and they'd stop you so you couldn't get a lamp to go down. That's how they laid you off. But then the unions came and they made all the difference in the world.

I am a union man, myself. They made me shop steward out at No. 8 Mine and I had to do my duty to the men. The company wanted to get rid of me because I was a union man. I think I was the only man on Vancouver Island who went that far. I was doing it for the good of the men. I still pay my dues – yeah, I stay with it.

I was in the mines for a long time, then I went and got an engineer's ticket and got out of the mines. I got all told a 50-year pin from the Union, a gold pin. – *George Edwards*

I was born in 1908 in Scotland. In 1921, I started in a coal mine on the picking table – I was only 13 years old. I worked there for one year and then I went underground.

You know, in the early days, you had to buy the fuel for your own lights. Working in the dark for the company, yet you had to buy your fuel for the light on your hat. You had to buy explosives. Oh yes, there were a lot of deductions before you got any wages. Wages were a big problem in the Cumberland area. It was as late as 1937 before we signed a contract with the United Mine Workers.

I worked at Cardiff, Alberta for a man who was reopening a small mine. There were five of us miners working there and the only way we could get paid was if the owner was able to sell the coal. To save money, three of us rented an old shack in Cardiff but it didn't have a stove in it. We heard about a stove for sale nearby. So, after work one night, me and my two partners went over there.

We knocked on the door and it was answered by a well-dressed woman. We said, "we hear you have a stove for sale." "Yes," she said, "come on in." The three of us went in and there were two other women about 65 years old. And while sitting and chatting, one said, "Would you like a drink?" Well, well, I hadn't drank but we said "sure." She poured us a whiskey and she started telling us about Cardiff. "Oh," she said, "this was a very busy town at one time. Yes," she said, "at one time I used to have as many as six girls." I thought you wouldn't need six servants in a house this size. I was that naive.

The last Chinese coal miner in Cumberland was Jumbo. He gave me his family's seal for helping him get relocated. He was very disillusioned because he was the last remaining Chinese coal miner in Cumberland. When he finally moved, he came to our house and cried.

Jumbo married when he was 16, before he came to Canada. When he came here, he had no children because he just married and left. Those days in China, your brother, your cousin or close friend could sire those children for you, and you were the Number One father. He had children in China, even though he wasn't the father. So he sent all his money back to China to support his wife and her children.

I am 74 years old. I think you're only as young as you think. I started out writing about my personal experiences and observations of my life in coal mining for my children but I wound up with a published book for others to read. It's the story of a breed of men, the pick miners, who are no more. I continue to write articles about coal mining for the local newspaper because it will all be forgotten if someone doesn't record it.

BILL JOHNSTONE

R E T I R E D   C O A L   M I N E R

JOE WHITE & BILL COTTLE

The first mine I worked in was the Atlas Mine near Drumheller. Drumheller was a going concern then. I worked seven years at the Atlas, ABC, Manitoban and Elgin Mines. I was 17 years old when I started.

When I came back to Nanaimo, I got a job at Extension No. 3, in the tunnel. During the Depression, I worked in the Jingle Pot Mine at Wellington and I also worked in No. 5 Mine in Cumberland.

I would say I had 30 to 40 years in the mines. It was good hard work. Oh yes, I would do it all over again if I had the opportunity. I am proud of the mines and the miners who worked them – there is no other group of workers in the world that made such a contribution to society. I would stand up in front of any group of men. I don't care who they are, I would stand and defend the coal miners of the world!

There are all kinds of old miners downtown – oh yeah, we meet in the mornings at the Safeway. – *Joe White*

My father was a coal miner but he got tangled up with the unions here. There was a big strike at No. 6 Mine (Wellington) and my father got blacklisted by the company for speaking out against the bosses.

When I was 14, I spent 10 years working the factory that made blasting powder in Nanaimo. The reason they put this plant here is when the first coal mines started, they had to bring in blasting powder all the way from England.

During conscription, I had wanted to enlist but I needed to work to help my parents. My father was disappointed that his boys didn't go into the coal mines. So when I had a chance to go underground at Lantzville, I did. I worked various mines including the Jingle Pot Mine, No. 1 Mine and the Granby Mine which was a hard-rock mine.

I didn't care too much for underground. I wanted to be on top.

My friend George Edwards said that after 75 years, you're living on borrowed time. I am 87 years old. – *Bill Cottle*

JACK RADFORD

My dad was a coal miner in Derbyshire, England and he ended up in Ladysmith where I was born in 1909. I had five brothers and two sisters – two brothers were killed in the mine. One was 19 and the other was 16.

In those days, when you left school, you went to work in the mines. I was 16 when I started at the Grandby Mine, six miles north of Ladysmith. My first job was working on the pit head, greasing cars. Then I went underground and ran an air winch.

I quit that mine when I was about 18 and went to Extension. In those days, there were about 600 miners living in Ladysmith and working at Extension and Grandby. Each day, we rode the train from Ladysmith to Extension.

One summer when I was 21 years old, I was riding the rails with some friends and we wound up in Drumheller. We had no trouble getting jobs at the Newcastle Mine. If you came from Vancouver Island, they hired you just like that.

I liked Drumheller. It was a good town for a young guy in them days – if you liked lots of fun. There were about 10 of us from the Island working there. Every payday, we would throw five dollars in the pot, go down and buy some liquor, and go to the dance at the Elks' Hall.

During the war, I worked in No. 10 Mine in South Wellington. Then I went into lumber because all the mines closed up and all the big logging companies came in. I spent 28 years in the mills as a lumber inspector. But I would sooner work in a coal mine than out in the woods.

This town is named after Ladysmith in South Africa. After the mines started to close, this place became a ghost town and you could buy any house for practically fifty dollars.

I keep busy all the time. I curl in the wintertime in Nanaimo. Summertime, I work around my yard – flowers and garden. I go downtown every morning for coffee with the boys.

MIKE RAINES

was born in 1907 on the Indian Reserve at Cedar, six miles from here. When we were young, we had only two choices: to the bush or to the mines. I was 15 years old when I went into a coal mine. I worked in the mines for the same company for 28 years. The mine was better than logging.

I dug coal and drove mule and that's practically everything. I was taken under a mule driver's wing when I went in the mine. He took a liking to me. He was a wonderful driver. I aspired to copy him and I made a wonderful driver myself.

I was a good union man, but I was never a lazy man. I was absolutely disappointed when they shut down the coal mines. You get bored with this hard life but I wish I had the time over again so I would never put in that dog time. I would be making it better.

I bought animals all my life. I had my first horse when I was 12 years old. I think the world of horses. I've been doing this stable business for 15 years. There's no money in it. It's just for keeping the animals. I have 40 some-odd horses.

JIM CRAIG & GRANDDAUGHTER, **MICHEL** TOBACCA

RETIRED COAL MINER

BILL MONCRIEF

I started coal mining in England, in Cumberland. My dad did everything in the mines – and he even played goal for Southampton.

I came to Canada when I was 23 or 24, and I brought my missus and two kids. I ended up in Nanaimo because my younger brother and brother-in-law were there, and I worked in a small mine there until it finished. In Cumberland, I worked in No. 4, No. 5, Scot Slope and No. 8 Mines.

Then I went to Tsable River Mine where I quit just before the mine shut down – I think it was in 1964. I liked all the mines but No. 8 Mine was a good mine.

I did everything: digging, loading, brushing, running a puncher and even the coal cutters.

But no, I wouldn't be bothered with coal anymore.

I've lived in this house 53 years. Hell, the old Chinese men used to come here and sell their vegetables. My missus liked the town and she didn't want to leave. Good enough for me – I was never one to want to move around.

My youngest son is the mayor of Cumberland. I can't give him any advice because I never see him now. It's a shame when you think of all the bloody coal they will never get out. There was a lot of coal wasted with poor mining.

I can't get around much any more because of my bloody legs. But Jesus Christ, if I could get around like I used to, I wouldn't care.

Of course, I used to love to go to Reno. Now I just sit here and play my music and enjoy myself. I am crazy on music. I listen to anything.

RED NICHOLAS & GRANDDAUGHTER, MICHEL DOUGALL

My father was a coal miner in England, Nova Scotia and Cumberland. I had an uncle working on the tipple in Cumberland and the boiler blew up and he got killed. Then, a couple of years later, my grandfather got killed in the mines in Nova Scotia.

I lived in Cumberland all my life. Cumberland used to be a city — it was the smallest city in Canada. We had 3,000 people in Chinatown. I'm going to tell you something: I have seen pigs being herded on this road and I'll bet you there were 150 to 200 pigs if there was one. And all the Chinese men, they would have their little switches on each side of them.

I worked 25 years in coal mines on Vancouver Island. I started when I was 14 years old at the No. 5 Mine in Cumberland. I started fire-bossing in early 1956.

During the war, I spent five years in the army. Then, right after the war, I came back to Cumberland and went back to work in the No. 5. I was single at the time and I should never have come back. Maybe, just to see my mom.

During the Depression, the British Columbia government brought down a law that said Orientals could not work in a mine. The reason they gave was that they were too dangerous, but the real reason was that there was a depression on and the Orientals could go to hell. It was a bunch of bullshit that they banned the Chinese!

There was a big fire here in 1935 that took out half of Chinatown. The company sold the remaining Chinese shacks for lumber at twenty-five or fifty dollars each. What they didn't tear down, they burnt down. I'm not kidding you.

I've seen guys come up for the last time, you know, come up from the shaft. The gates clunk behind them. Fifty years of work and they just say, "Thank you very much, goodbye." That sort of thing. I've seen it happen.

GEORGE LYONS

I was born in 1905 in Wales and I came to Canada with my father, a coal miner. He took a job at the mine at Coal Creek near Fernie.

I started working for the coal company at Fernie when I was 16, repairing mine cars in the repair shop. During a nine-month strike in 1924, I went to Kimberley and worked underground. After the strike, I loaded ashes in the boiler room. This job led me to get my first-class engineer's ticket.

I came over to Cumberland in 1929 and got a job at No. 4 Mine as a timberman's helper. They also stuck me on driving mules. I stayed in the No. 4 Mine until it closed, then moved to No. 5 Mine where my next job was digging coal and running mining machines on the 'longwall'.

I never felt better in my life than when I was digging coal, though I weighed only 145 pounds.

Then I went to No. 8 mine and stayed to the end of the mine in 1955.

In 1954, I discovered I had silicosis.

My last years in coal mining were at Tsable River where I was an underground hoist man. Once you get into coal mining, something gets into you and you stay with it. You get to like your work. You are on your own when you're underground mining.

LEN COOPER

hen I was 13 years old, my first job was in a carpentry shop in south Yorkshire. I had to wait until I was 14 to go underground. The day before I was 14, I asked the mine manager if I could work underground. "Fourteen! Yup," he said, start tomorrow. I went home and told my mother and she started crying.

My older brother decided we should move to Canada. I wanted to go to Australia but he decided for us. We ended up in Nanaimo, where I worked in the Protection and No. 1 Mines from 1927 to 1937.

At the end, the Protection Mine was pretty well worked out. Sea water was forever coming into the mine and they couldn't stop it. You got soaking wet while you worked. You could even hear the turbines of the damn boats while you were working underground. I came to Cumberland from Nanaimo. The first place I went was the No. 5 mine. When I got there,

I found that the manager from Protection had been manager at No. 5, so he gave me a job right away. I was a pony driver, then I was digging coal.

But I got no money, I got no clothes. One fellow said I could use his son's clothes because his son was hurt and off work.

When I first came to Cumberland, I stayed in the Union Hotel for thirty-two dollars a week. It was a boarding house for miners. There was a beer parlor in the hotel. Later, my wife and I used to go there and have four 10-cent glasses of beer.

I had my fire boss papers from England and Nanaimo. I became a spare fire boss at No. 5 Mine and I did that for a while. I worked 51 years in coal mining. I had good luck. I never got hurt.

Oh yeah, I'd do it over again. Most miners are good guys to work with. Hey, if you need a hand, they'll be right over to help you. No matter what you are doing, you get help.

Coal mining was a job and I liked it. I worked in the bush three years and I think the bush was every bit as dangerous as the mine. I always said you could get killed walking across the street. Down in the mine, you were aware of the danger and took care of it. Accidents were mostly human error and carelessness.

I was born in the old country, in England. We moved from England to Glace Bay because I had an uncle there. Then that same uncle moved out here and we followed him. We moved to Bevan and my father got a job in the lamp cabin at No. 7 Mine.

I was 14 years old in 1915 when I began work in the mine. I started to work in the lamp cabin at $1.25 a day. When the war ended in 1918, the company gave returning soldiers jobs at the mine. Two ex-soldiers got jobs in the lamp cabin.

So my dad had to go on the picking tables and I went into the mine. In No. 7 Mine, I operated various small hoists and a battery locomotive. When the No. 7 shut down in 1921, I was moved to No. 4 Mine and did rope-riding. When I was a faceman in No. 5, looking after the pans, I wrote and received my fire boss ticket. So then when I started at No. 8 Mine in 1937, I was a fire boss.

I fire-bossed until 1953 when they closed down the No. 8 – I was the last man out of the mine.

I worked four years as fire boss on night shift until finally I went to the boss and told him I wanted off night shift. "Oh no," he said, "I can't take you off night shift." "Well," I said, "you might as well write my time out because I can't sleep. I can't look at the kids anymore because I jump down their throats, I'm that short-tempered."

I worked for one man who used to get so mad that he would throw his hat down on the ground and jump on it. He was an awful man who would fire you and then, when you started to go home, would ask where the hell you were going. He played the organ in church but he was the foulest-mouth man you ever did meet. He'd get mad at you and bawl you out and then immediately forget about it. If you didn't speak to him the next day, he would say, "What the hell is the matter with you?"

I played soccer for the Eagles. We played together on a team in 1923 when we went to Vancouver and won the B.C. Championship. There are only about five guys I played soccer with who are still alive.

These days, I do a lot of reading. I can't go into large crowds because of my hearing handicap. All I hear is a loud noise.

JIMMY WEIR

GEORGE MARSHAL & BILL MARSHAL

I was born in 1905 in Scotland. My father was a miner. I was a miner. All the family were miners.

When I was 13 years old, I started as a miner in the old country, working in a shale mine 12 miles outside Edinburgh. They mined the shale the same way you mine coal.

We came to Bevan, near Cumberland, in 1921 because of the coal mines. My first job was working as a coupler for the mine cars.

The Bevan Mine was just shutting down when we arrived so I got a job in the Cumberland No. 4 Mine. I was in No. 4 for 14 years, then it shut down. So I went to No. 5 Mine. When it shut down, I went to the Tsable River Mine. I spent four years at Tsable River.

Coal mining was all we knew. It was fine with us. We didn't know any better. – *George Marshal*

I play the accordion. I play for my own enjoyment unless somebody wants a tune. Then I play for them.

I was born in 1900, in Scotland and I worked underground in the shale mine there from when I was 14 to 20 years old. I drove a horse.

In Cumberland, I never dug coal – I worked on the haulage all the time. Then I was a rope-splicer for 25 years. I retired when the mine shut down in 1966.

I guess I would still be working if the mines hadn't stopped. I got 52 years, counting the old country.

It used to be that I had cigars in this pocket, a pipe, pipe tobacco and matches in that pocket, cigarettes in this pocket, and a tin of tobacco in this pocket to roll more. I was scared I would leave home and forget my tobacco – really scared that I was going to find myself 10 miles down the road and find I hadn't got my tobacco. And in the end, that's what made me stop. Really stupid, isn't it? – *Bill Marshal*

JIMMY COCHRANE

I was 15 when I first went underground at the Protection Mine in Nanaimo. You used to have to get on a scow to take you across the water to the other side. As soon as the scow landed, all the men and boys would run up the hill like a bat out of heck to get on the cage to get down. They were just dying to get to work.

It was hard work but a certain amount of fun. You knew everybody. There was one thing about working in a coal mine: you would never see too much fighting or scrapping. No fisticuffs. You could get as mad as you wanted at the fellow alongside of you but you still had to work with him.

I have pretty near 50 years in the various mines in Nanaimo and Cumberland. I started in No. 5 Mine and stayed until it closed down. Then I shut down No. 8 Mine in Cumberland and finished up in Tsable River Mine. I closed them all and worked myself out of a job.

The mine manager at No. 5 Mine encouraged me to be a fire boss. He always said you were wasting your time shovelling coal. I became a fire boss because I was always looking for that extra dollar. I don't know why because I would always spend it just as fast.

I was the manager at Tsable River. I was an outspoken so-and-so. They didn't like it when I spoke my piece.

When the coal business closed in Cumberland, I got a job in a machine shop in Courtney for about 10 years. I retired three years ago because I couldn't read my own writing on the cheques I was signing.

I got asthma when I was 65. Up until then I had never been sick. I am 73 now and all I can manage is a little golf.

MINE OFFICIALS

DARRELL HICKS

TIPPLE BUILDINGS

LLOYD GETHING

started coal mining in 1937. My family owned coal mines in the Peace River area, near Hudson Hope. But before I joined the coal mining business, I taught school for eight years.

We closed down in the Peace River area in the spring of '51. We'd run into financial difficulty because there were no markets for coal. My brother, King, continued to operate the King Gething Mine until the late '60s.

In 1951, I started as an underground miner at Telkwa and gradually worked my way up to superintendent of the mines. I'm now part owner of a coal property near Telkwa and I'm trying to develop a new surface mine for the export market.

It's a very calm sort of life, you know, coal mining. It's different than going into a hard rock mine where you're clambering up and down dangerous places, trying to follow a lead that may not be there. But in coal mining, you're using your knowledge of what coal seams do. At all times, you use your knowledge of rock structure, right from the first day you start mining.

That's one of the attractions of coal mining. You've likely sensed that coal miners like to mine coal. Long after they've retired, they're still talking about it but with a feeling of having enjoyed it.

Here's another interesting fact: if you check the statistics of how long people live, the worker who lives longest in Canada is called a farmer. The second-longest-living worker in Canada is called a coal miner. Everyone knows that the reason a farmer lives a long life is that it's a calm existence. He's using his head, he's running his own working place. Where he plants his carrots is his own working place.

And in coal mining, you also — sort of — run your own working place. It's a calm life.

KEN MOORE & **BRIAN** BRICKER

FRANK SALT

KEN INKPEN

DOUG NEWHOOK & **STAN** HOWARD

DAVID MORGAN

DOUG ENNIS

MINE BUILDING FOUNDATIONS

JAMES MOODY

M y father came to Canada in 1910 from Dundee, Scotland – he was a coal miner. I followed him a year later, when I was 17.

When I got a job, it was at the Middlesboro Colliery near Merritt. I was in the mine from 1911 to 1914, then I went into the army. When I came back, I started in the mine again and stayed until it closed in 1940.

My first job was working as a mule-skinner. We took the empties into the mine and brought out the loads to the tipple. We had a three-mule team – only one horse, the rest were mules. You had to be a good swearer to drive those mules, you didn't mince your words with them. They're stubborn, very stubborn.

I also had a job bucking chutes. That was in 1912 when I was making $2.75 a day. In those days, you didn't have a lamp on your head – you carried a safety lamp hanging on your belt or hanging around your neck. Some of the chutes were low and you had to lay on your back in places. That lamp used to sit on your chest under your chin and it got pretty hot.

When I came back after the war, I was a miner on the coal face. I was getting the big wages then: $3.30 a day. Those were top wages for a coal miner.

Most of the miners at the Middlesboro Colliery were of Scottish origin. I had two brothers who worked in the mines here – I think the owner brought Scots over just to mine coal.

Well, it was a job for us. You know, we worked maybe two or three days a week. They'd tell you when you came off shift that if the whistle don't blow, then you come to work. But if it blows, stay home.

I was always a common coal miner. I was used to the mine and I never knew anything else. But it was pretty hard to raise a family of four on only two or three days of work a week.

TIPPLE & WASHPLANT BUILDING

RALPH LARNER

My dad was always a coal miner — he mined a coal seam 20 inches high in Staffordshire. We came to Canada in 1906 because he'd heard the streets were paved with gold.

In 1910, when I was 14, I started at the No. 2 Mine at Coal Creek. All I had to do was open and shut a door to let the horses and mine cars come through.

My next job was bell-rapping to make the hoist go up and down. I did that for about four years.

I drove horses and became so good at it that the mine gave me a special job of breaking in the horses. They used to bring in prairie horses that were like lions — they took a lot of breaking in. Then I had a job as a rope-splicer. After rope-splicing for about five years, I started mining coal in the No. 1 East Mine. I sat for my fire boss' papers. I kept digging coal, waiting for a fire boss job to open.

I ended up fire-bossing in Coal Creek for 17 years. When I was a fire boss, I often would pick up a shovel and help the miners. When Coal Creek closed, I was offered a job at Michel or had a chance at a pension.

I worked 47 years in Coal Creek mines. After retiring from the mines, I got tired of sitting around, so I went to work for eight years at the Fernie Hospital as the boiler man.

To tell you the truth, even though it was hard work, I enjoyed it. It was dangerous work. I had three broken legs. I think, when working in the dark, the time passes quicker. For me it did.

I was one of the original founders of the Rod and Gun Club in Fernie. The old Central Hotel was a popular place for coal miners because it had the best beer.

Fernie has had its up and downs — its fires, floods and strikes. But Fernie always had the name as a friendly town.

RETIRED FIRE BOSS

I was born in England in 1913, and came to Canada in 1927. I went to school for a little while, then finally went to the mines. I started in the '30s when there wasn't much else to do here – mining was the only thing.

I started in No. 1 East Mine as a boy driver and worked my way up. In those years, No. 1 East Mine had a lot of bumps – I was right in most of them myself. You could feel the bumps on surface and sometimes as far away as Fernie. I guess that mine might have been one of the worst in the world for bumps and for men getting killed. When we felt the house shake, we would head in. Get a lamp and head into the mine to help out. We wouldn't need to be asked, we would just go.

In 1939, I got hurt when an 800-pound rock fell on my back. It took six men to lift one side of the rock so they could get me out. Although I was somewhat paralyzed, I could stand and walk out of the mine. My back hurt me, but I went to work again.

Then in 1946, my partner and me were hit with about two tons of coal. The coal just exploded due to rock and gas pressure. We walked out of that one.

Two years later, they put a steel plate in my back and I was off work a year and a half. After that, I worked in the mine another 10 years, doing mechanical work and rope-splicing.

All in all, I kinda enjoyed mining. The people I worked with were mostly good people just striving for a living. – *Bill Chester*

I came from a mining town in England called Saint Helen's – we had a nice home there. Then I came here in 1923, with my parents. My mother never did adapt to the mining town life with its outdoor privies.

We came from England in October. I remember because the kids made fun of us – my brother had on his short pants and his knees were freezing. At the first snowfall, we each got a toque and a pair of mitts from the store. I think my little brother even got a sleigh. They gave us a party every Christmas. Every child up to 14 years of age got a nice gift. I remember my first gift was an Eaton's beauty doll. I just grabbed it and headed home. It was so beautiful.

It was a good life in a lot of ways. Living in Coal Creek and Fernie, you never thought of discrimination. There was every nationality here and we all grew up as family. You might hear the odd word – "bullhunk" or something – but it didn't mean a damn thing. When it came down to the nitty-gritty, you thought the world of one another. – *Jane Chester*

JANE AND BILL CHESTER

Being a mine manager was the best job because you were totally in charge. Well, coal mining had its good moments and its bad moments. Generally speaking, I would say I enjoyed it and never regretted it for a minute.

I was born in Wales, in 1906. I started mining in 1920 at the age of 14 in the No. 4 Mine in Cumberland on Vancouver Island.

I was trapping doors on my first day. This driver boss came along and said to me, "So you are Morgan." I said, "Yes sir!" He said, "Well, look, we have a little initiation before we start."

So they grabbed a hold of me – they used to have this slimy oil and grease they used for greasing rollers and wheels for the ropes to run on – they opened the back of my shirt and poured that mixture down my back and then three or four handfuls of sand and gave me a kick in the butt and said, "So, okay! Now you're away!"

That was the start.

I worked in Cumberland six or seven years, then I went logging. Then I worked in an iron works. In 1932, I went back to Cumberland and started in the mines again. I got my fire boss papers and my pit boss papers. In 1942, I came up here to work in Michel as pit boss in the No. 5 Seam Mine.

In 1965, I got my mine manager's papers. I retired seven years later. I started golfing at Comox while I was working in Cumberland and have played continually from 1954. My handicap is 10 or 12. I had a 38 here (at Fernie) yesterday and I've shot my age twice this year.

In order to beat this game, you have to hit the ball straight down the centre. – *Irvine Morgan*

I was born in Coal Creek, in 1910. My father was a coal miner for 50 years and both my grandfathers were coal miners. When I was two years old, we went back to England. I left school at 14 and, at 15, I went to work on the pit top at Harrington Colliery in Cumberland. As soon as I was 16, I went underground. I worked there until we came back to Canada and Coal Creek in 1927.

I used to read my dad's coal miner's pocketbook when I first started work. I failed my first fire boss' exam mainly because of my English – I couldn't put stuff on paper. I went to night school and finally got a teacher who could teach English as well as math.

When I got my fire boss ticket, it was the beginning of a new era. I was the first young official in the late 1930s – the first of a new generation. There were about 20 years difference between me and the next official above me. During the Depression, there were no new men, no new blood.

I was a dyed-in-the-wool miner. I just enjoyed the challenge of trying to beat the physical aspects of running a coal mine – the things you run into and have to overcome between the gas and water pressure. Then, on the other side of it, you've got the men to handle.

In 1963, I lost my leg when we had a cave-in and it pushed me under the ripper bar on the mining machine. The bar cut my leg right off through the knee. I could see my boot with the other part of my leg lying near me.

The operator of the machine probably saved my life by dropping the ripper bar and shutting down the machine. Being the official down there, I took charge because the other men were at a loss. I told them to put a tourniquet on me and I told myself I had to fight to stay alive. I remained conscious until I went through the hospital door.

After I lost my leg, I got an artificial leg and used golf as therapy. For the first ten years, I walked the course with my artificial leg. It's only the last six years that I've been using a cart. – *Dan Chester*

IRVINE MORGAN & DAN CHESTER

ARNOLD WEBSTER

I don't think anything would be really worthwhile if there were never any problems attached to it. The last seven or eight years before I retired, I was the underground safety co-ordinator at Michel. Being the safety co-ordinator produced the greatest number of headaches. It's hard to be safe for someone else. The biggest part of the job is going underground, observing conditions and trying to correct the men who were doing things the wrong way.

I was 17 when I started in Coal Creek as a boy driver. In those days, you worked only one or two days a week — I was paid $3.15 a day. In 1936, you took any work that was available and, around here, there was nothing but coal mines.

Then, after the war, mining picked up. I was in the army and came back home and went right into the mine. You know,

coal has always had a history of boom and bust, but I was fortunate in that after I started working, I was never out of a job.

I started fire-bossing in 1950 and by 1952, I was pit boss at Coal Creek. In 1958, the Coal Creek Mines shut down altogether. All in all, coal mining has provided me with a reasonably good life. My wife, Sheila, and I got married, we had three children and raised and educated them. One of my two sons is the principal of a school in Fernie and the other works in industrial relations for B.C. Coal (Westar). All three kids were raised on a coal miner's salary.

As long as I can remember, the men who worked underground have had a lot of pride. To us underground types, there is only one true coal miner and that's a guy underground. The other guys are producing coal, but...

JOE JOSS

I was born in Burmis, Alberta. When I first came to British Columbia, I worked in the mine at Blakeburn. I was 19 years old. That was the first day I dug coal.

Then I went to Michel, the old No. 5. After that, I worked in the Blairmore Mines for eight years. At the start, my jobs were driving horses and rope-riding. Then I got my ticket and went diggin' coal.

At Blairmore, I was on contract. I made good money there – fifty-eight dollars a day in 1952. You blast one shot and boy! the coal would come down. The seam was up to 80 feet high.

After Blairmore, I came back to Michel as a carpenter for 10 years. Then I quit the carpentry business and got on the electrical crew. I have done everything: carpenter, welder, electrician. I can do any goddamn thing. I got a job in the electrical group because I learn fast. I said to the electrical boss, "Why did you hire me? I don't know nothing about electrical."

He said, "You be here one year with me, Joe, and you'll know everything." By Jesus, I'll tell you, he sure trained me fast. He was a rough man, mind you. I said, "Listen here, Jeff," and I put my hand on his shoulder, "If I need shit, you give me shit." He said, "I'll give you shit if you need it." But he didn't do nothing.

In Michel, when there wasn't much work, I used to make things for myself. I made knives for cutting leather. I made belts and purses and everything. Now I even carve saddles, but I do it at home. Here in my shop, I always get interrupted.

I opened up this shoe repair shop eight years ago and then I quit working in the mine. I had to quit. But I was only 60 and too young to retire.

I have had a lot of jobs – I go steady. I'll tell you, a lot of times if I got two hours of sleep I was okay. Yessir, yessir, that is no bullshit.

BILL BROWN

Coal mining is in my blood and I'm glad I never quit mining because now I have early retirement. If I had my life to live over – I've told different guys this – if I had my life to live over, I would never work so hard. I did every job in the mine and we used to work six days a week.

My dad had a fourth-class boiler ticket. He worked in the boiler room at Coal Creek. There were seven of us in the family and we were all born in Coal Creek. Four of us worked in the mines.

When I was 19 and working in No. 1 East Mine in Coal Creek, my partner was killed by a bump that hit us just as we were ready to go home. I have always said that if you're going to get killed, you're going to get killed, no matter where you are. I think the alarm clock is set for you.

I was never scared in the mine. If you're scared, it's no use working that job. One of my brothers was in the Balmer explosion. He never got hurt too bad but he died of cancer a year ago, and he was only 52. That bloody cancer is a son-of-a-gun, you know.

I am 58 years old now. I quit because they gave me a good pension. Don't you think 43 years is long enough? If us old guys didn't quit, the younger guys would get laid off.

I go fishing and go out for walks. I have been busy helping people since I retired. Not for money – maybe just a meal. To me, an apartment is the best and I got a Dodge truck.

JAZZ ANDERSON

I was born in Mountain Park, Alberta, in 1920. My dad was a coal miner and mine owner at Mountain Park. He was killed in Mountain Park, in 1922.

I started working at Coal Creek when I was 19 years old as a loader on shaker conveyors in No. 1 East. I worked at Coal Creek from 1939 to 1958 when the underground mine closed down.

My brother got killed in No. 1 East on his last day of work. So I stayed and looked after my mother. I never did go to college but I did my own studies. I was fortunate because of all the old mining books that were around the house. So I got right into it. I made a hobby of buying mining books, up-to-date mining books. I still keep right up-to-date on the latest issues in mining – I spend four hours a day reading new text books.

When Coal Creek closed, I went to Michel and helped open a new mine, No. 8 South. There we started to use continuous miners and got electrical and diesel equipment into the mine.

At Michel, I received my first-class certificate and became the combined manager in charge of the whole area. I got inspiration to advance myself because I was brought up in a mining area.

During my career in coal mining, I have opened up a total of 22 underground coal mines plus 11 strip mines and three preparation plants. I have lost five members of my family in coal mines.

I was always stimulated by the idea that a coal mine could be a safe place to work. You can make them safe and economical, I know. I have travelled to most coal mining areas – I'm completely familiar with coal mining. You name it, I've done it, I've done practically everything in coal mining. I worked in coal mining 43 years. I've been associated with everybody in the industry, I knew them all.

I am a full member of the Internal Order of the Black Hat.

MIKE CAMILLI

I was born in Fernie in 1916 and started school in Hosmer in 1928. My father was a coal miner and a Great Northern section foreman here in Hosmer. When I was a kid, we used to get a basketful of fish out of the Hosmer Creek.

In 1934, when I was 17 or 18 years old, I was an accordion player in a band in Hosmer. My uncle was a fire boss at Nordegg and also an accordion player. He asked me to come up to Nordegg to play in their band. I really had nothing to do so I went up there and got into the band.

Then I got into the mine. All the people were from the Fernie area. In 1938, when they started to lay off men at Nordegg, I came back to Hosmer because my family was here. The Hosmer Mine was closed so I got a job at Michel in the mine.

I spent 31 years in the coal mines. Then I went into the bush and worked in a mill until they retired me.

I played in bands for more than 50 years but I don't do it anymore. Now all I do is play for seniors in the old folks' home. When I was in my prime, they told me I was the best accordion player in the east Kootenays — that's what they told me. I'm rusty. I don't play anymore. I used to play every night, no matter what.

In the old days, nobody was working. We used to have what we called 'unemployed dances.' You could get into the hall for 10 cents and then you played whist. You could win a sack of flour or a ton of coal, something like that for 10 cents. Then you'd get two hours of dancing. The band would play and we'd each get two dollars — also a cup of coffee and a sandwich. Everybody in Fernie would be there.

JOHN BACHLET

I was born in 1901 in Czechoslovakia. I first started coal mining at the Hillcrest Mine in Alberta in 1926. Then I came to Fernie in 1928 and got a job in the mine at Coal Creek. I also worked four or five years at Michel.

When I started, it was strictly hand-loading with shovel and pick. I quit in 1963.

Now I keep busy with rabbits and chickens. I have 29 rabbits – they are a Belgian breed. In the old days, I always had rabbits. I have a big yard to grow hay. There is always something to do when you have a big yard.

I liked digging coal the best. You betcha your life I liked being a coal miner – I was lucky! Twice, I was a little bit hurt, but not that bad.

One time, my boss says, "Hey, John, you work a long time in the mine. You want I should put you outside? Take you from inside and put you outside?" I say, "No, thanks" and he just looks at me. I say, "What's the matter now?" He says, "John, if you like the job, it's pleasure. If you don't, it's murder."

JOE PARKER

oal Creek used to be a beautiful spot but now it's all grown in. There were lots of houses and even a school. My daughter was born here. I lived across the creek from that wooden bridge.

I was born in 1907 in Coal Creek and I was 16 years old when I started work here. My job was helping the track layer. Then they gave me a job rope-riding. I was about 20 when I started rope-riding, and I did that for years.

I liked to be energetic – running around and getting on the rope and ringing the bells. In 1940, I got on the hoist and I stayed on the hoist. The big underground hoist. I liked rope-riding better than being a hoist man. It took me a long time to get used to sitting on the hoist and watching the drum all the time. The only thing that made me stay on was that little extra money.

When they closed Coal Creek, I went down to Michel. Our crew was one of the first to start hydraulic mining. At Coal Creek, it used to bump a little and some of the guys would run, but eventually you got used to it. I always said, if you could feel the bump, you were all right. It was when you didn't feel the bump that there was a problem.

When we used to live up on the mountain side of Coal Creek, we would be sitting at the kitchen table and everything would start to shake on the table. Then you knew that there was a big bump in the mine.

In 1976, I lost a son in the Balmer North Mine explosion. He was 28 years old. I worked the day shift in the same mine. I was in the wash house when we heard that there was something wrong at the beginning of the next shift.

Balmer North was loaded with gas – I didn't like it, myself. You could hear the gas coming out of the coal. It's a wonder it didn't blow up before it did.

Natal Railway Station, 1982

Natal Railway Station, 1986

ELECTRICAL SHOP BUILDING

BUSTER DUFORD

M y father was quite a character. When he died in 1964, he was the second-longest-living resident of Fernie. He was 75 years old. He came to this part of the country when the CPR laid the steel here in 1898 – that's when coal mining first started.

My father came from Canmore – his father was a blacksmith at Canmore Mines. My father was a railroad man for 50 years. He became superintendent of the five-mile-long railway that the Crows Nest Pass Coal Company operated from Fernie to the mines at Coal Creek. He always used to say "It's not long as the CPR but it's just as wide."

I was born and raised in Fernie and spent the first five years after high school working in a hardware store. In 1949, I was transferred to Michel to manage a hardware store for eighty-five dollars a month. I spent one year in that store and then discovered I could make more money as a labourer for the Crows Nest Pass Coal Company at the coke ovens. At the time, they paid about $4.90 a day, six days a week.

I started out as a labourer and, inside of two years, they made me plant superintendent. Then they moved me up to the surface mine where I spent five years. Then they moved me back to the coke ovens in 1955 and I stayed there until 1970.

By then, I'd had enough. My health wasn't too good. It was a dirty place to work.

Before my official retirement in 1980, I worked in safety, hiring and community relations. These days, I'm working on an hourly contract to take visitors around the property. Last year, I conducted 147 tours of the property and looked after 1,400 people.

In the 27 years I worked with the coke ovens, over 800 people worked there. Once they shut down the ovens, they sure deteriorated – everything is getting rusty. It's just two years since they closed down the Byproducts Plant and look how it's all starting to grow over.

WESTAR COAL MINERS

Curt Bellerose, Keith Bracewell, George Bradbury, Anjon Chowdhury,
Bob Desjordins, Eddy Egom, Clive Endicott, Jerry Facette, Charlie Fanshaw,
Dave Hudson, Stan Janson, Don Nastas, Matt Nijam, Doug Peck,
Chester Taje, Bubba Vanloon, Alan Wheadon, Bob White

HARVEY TRAVIS, OVERMAN & **PHIL** BORTNIK, MAINTENANCE SUPERINTENDENT

ROBY SLOPAK & **WAYNE** HALLADAY

NICK SCIARRA

BOB HUTTON, **TONY** ZIELINSKI, **PAUL** STOREY
& **CHARLIE** TESSMAN

JOE REPKA, WELDER; **J.** POLACIK, GENERAL FOREMAN;
V. KIBALA, MECHANIC; **L.** TEGLAS, MECHANIC; **M.** MEAKIN, WELDER;
ALF BERNARDO, MECHANIC & **A.** GUIMONT, WELDER

GERRY DESSERRE, WELDER; **NORRIS** FEDOREK, MAINTENANCE PLANNER;
& **DON** CURRY, MAINTENANCE FOREMAN

KEN FRASER, **RON** PORTER & **AL** BLANCHETTE

RON MICHAUD

JANICE HEIGHEY

STAN KUTA

JOHN KINNEAR

MACHINE SHOP

RODDY BLACKJACK

I was born in 1927 at Five Fingers in the Yukon. My father was a trapper. I started working on the steamboats on the Yukon River when I was 15 years old.

In 1947, I started at the Tantulus Butte Mine in Carmacks. My first job was loading coal onto river barges. Then I ran the air compressor to supply power to the underground machines. When anything went wrong, I had to go underground to help repair it.

In later years, I ran the bus and drove a truck for the mine. I finished in 1974.

Us guys were good coal miners.

MORRIS HARDING

I was born in County Durham, England, in 1922. My grandfather, father and my six brothers worked in the mines in England. I started in the Holden Colliery, 14 miles south of Newcastle, when I was 14 years old. My first job was trapping the ventilation doors for ponies. During my years there, I held jobs as miner, labour foreman, fire boss and mine foreman.

After 34 years in the same mine in England, I came to realize I was sick of mining, so I emigrated to Vancouver to help my brother at his gas station. That was the worst bloody job I ever had. I didn't like Vancouver because of the rain. So I said to the wife, "Let's go back to England" but I didn't want to be in the mines again.

In 1971, I heard they were looking for men in Carmacks. I said, "Where the hell is the Yukon?" I finally decided I might as well freeze in the Yukon, so I took a job as fire boss at the Tantulus Butte Coal Mine. When I came to Carmacks, what shocked me most was the timbering system. Here, the coal seam is on a 30-degree angle and I was used to flat seams. You had to climb all these ladders and it was nerve-wracking — especially watching the native Indian coal miners mine a 20-foot-thick seam.

I was at the mine six months before the native Indian coal miners would have anything to do with me. After they found out I was here to stay, they accepted me. Most of them were good workers. I held the job of fire boss until the mine caught fire in 1976. It started in an old mined-out area. After the fire, it was too bloody dangerous to start the mine again, so I worked at the Tantulus surface mine until 1982 when I took early retirement.

I suppose you get used to one job — that's why I came back to the mines. I figure I got about 45 years of coal-mining experience. If you want something better, you have to take schooling.

CLYDE BLACKJACK

I was about 15 years old when I first started working summer holidays at the coal mine. My first job was sacking coal. We sometimes had to sort through the lump coal and throw away the rock. I was getting $1.25 an hour.

I was sacking coal for about three years before they finally put me on the mine trammers. I trammed coal out from underground. After two years, I climbed up to miner's helper. I remained miner's helper for five years until they finally said I could be a miner. From then on, I did mining, tunnelling, rock drilling and everything. Most of the time, I operated a raise-mining machine. I would go so far out I would come out on surface in the bushes and could snare a rabbit.

I left the mine in 1971 before they shut it down. I'm glad I left because now I have a better job: I am chief of the Little Salmon-Carmacks Indian Band. Now I have to work full-time as chief for the people because it is a full-time job. There are some things I have to look after: education, social programs, administration and land claims. The land claims are really full-time work now.

I think of Carmacks as a coal mining town. Coal mining has taken place here since the 1800s. My dad and my grandfather worked in the mine. The mine on the other side of the river, the Tantulus Mine, operated before I was born – about 1902 to 1910. My grandfather would have worked in that mine.

Us guys from Carmacks have some good experience and we know how to mine that coal. I liked to work in the mine – it was very good experience. It was nice to be working closely with my people.

A new coal mine in Carmacks would provide jobs for my people. It would certainly help. Something has got to be done.

BILL CASHIN

I was born in 1927, in Eastern Passage, Nova Scotia. When I was 16, I was working at the shipyards in Dartmouth. I heard they were looking for a crew for the St. Roch, the RCMP boat, so I joined on.

I was a seaman and a bosun on that historic voyage. We travelled by boat from the east coast to the west coast through the Northwest Passage. I was just a kid.

I became a special constable for the RCMP. The next few years, the St. Roche made other trips into the north to supply RCMP posts. I first visited the Yukon and the Northwest Territories at Herschel Island. There were only five of us crew members still alive in 1974 when they put the St. Roch into a Vancouver boat museum.

I liked the north. After that voyage, I took a job at Keno Hill Mines where I worked in the mill and underground. In 1955, I moved from Keno Hill to Carmacks. I started out cutting mining timber.

In 1962, I started work in the Tantulus Butte Coal Mine, first as a miner's helper. Then, for ten years, I was fire boss at the mine. I looked after the mine by myself for a few years.

The Tantulus Butte Mine caught fire in 1976 and is still burning. Everybody used to smoke underground. In recent years, we mined some coal on surface, but the Tantulus was an underground mine, with the main seam standing almost vertical. We mined the coal like it was a hard-rock mine. Throughout the mine were these shaky wooden ladders – the native coal miners would run up and down these ladders with no problem.

I really liked mining. You'd put in your eight hours and that was it. I enjoy everything I do but the last three years here have been difficult. We've been waiting to see if the company will reopen the mine.

Coal Miners

of ALBERTA

Alberta

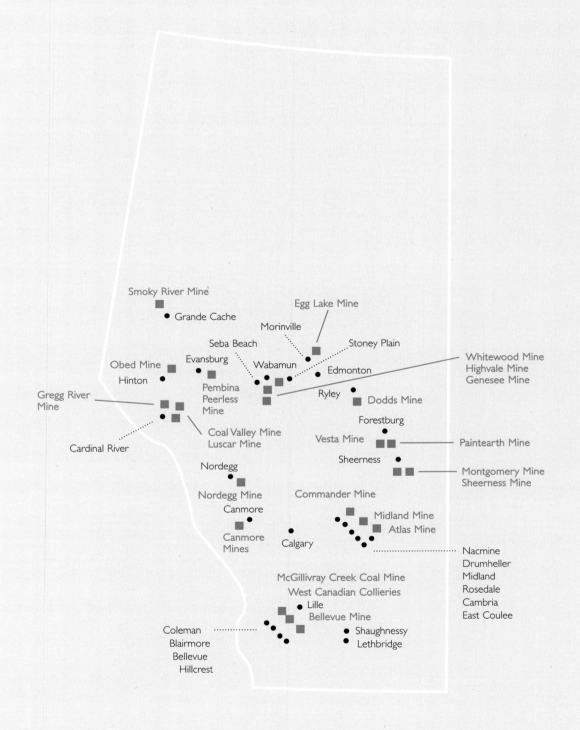

Smoky River Mine
● Grande Cache

Egg Lake Mine

Morinville

Seba Beach   Stoney Plain

Evansburg   Wabamun

Obed Mine   Edmonton   Whitewood Mine
Hinton   Highvale Mine
                                                          Genesee Mine

Pembina   Ryley
Gregg River   Peerless   Dodds Mine
Mine   Mine

                                  Forestburg

Coal Valley Mine   Vesta Mine   Paintearth Mine
Cardinal River   Luscar Mine
                                  Sheerness

                                                          Montgomery Mine
Nordegg   Sheerness Mine

Nordegg Mine   Commander Mine
Canmore
                              Midland Mine
                              Atlas Mine
Canmore
Mines   Calgary
                                                    Nacmine
                                                    Drumheller
McGillivray Creek Coal Mine   Midland
West Canadian Collieries   Rosedale
                                                    Cambria
                              ● Lille   East Coulee
                              Bellevue Mine
Coleman
Blairmore   ● Shaughnessy
Bellevue   ● Lethbridge
Hillcrest

● PHOTOGRAPH LOCATION   ■ COAL MINE LOCATION

JACK MARCONI

I was born here in 1915 and lived here all my life. My father was Italian and my mother, English. My father worked in Lille first and then moved to Coleman about 1910.

When I started in the International Mine in 1930, I was earning $3.50 a day. I got my fire boss papers in 1941 and, in 1944, I got my pit boss papers. I was the first pit boss at Vicary Mine and I was pit boss there when I retired.

I retired in 1973. I stayed underground 43 years and I didn't mind it except when I was 15 years old. I didn't like working in the mine then, maybe because I was scared.

After I retired, I stayed involved in the mine rescue team. We won a number of awards, including both the provincial and Dominion championships.

We were very unhappy about the coal dust from the tipple blowing into our houses and everywhere else in Coleman. But when I worked for the mine, I couldn't say anything about the coal dust because I was an official.

Now that I'm retired, I have complained many times to the company and the government officials.

FERUCIO DeCECCO

If my father had been alive, he would have broken both my legs before he let me go underground. But I didn't mind it. Actually, there was a lot of friendship and fun down the mine. I think that's the one thing that's really missing now in any job.

My father came here from Italy in 1903, the year of the Frank Slide. He worked at the Leitch Collieries as a stone mason. I was born in 1917 in Coleman. I had to quit school to go to work and the only thing there was around here was a job in the mines.

I had just turned 17 and had to lie about my age to get into the International Mine. My first job, at $3.50 a day, was trapper, opening doors. This increased to $4.45 for bucking coal. Then I drove the dinky locomotive. I rode rope and I packed timbers until finally I got my miner's papers. After the mine, I did various jobs. Then I started a little confectionery store.

There were a lot of Italian, Slovak and Polish people in Coleman who came to work in the mines. I think there was a lot of resentment because most of the good jobs were held by English-speaking people.

You can understand that, because most of the old-country people who came here didn't have the education for a job like time-keeper.

JOSEPHINE AND BILL MISZTAL

Oh, I remember those days. There were lots of ambulances going to the mine, lots of accidents. I didn't like Bill working underground. There was too much gas in the mine. It seemed that every day, there was another accident. I'd just wait for four o'clock when the bus would come. You'd worry that somebody might phone you to say he's in the hospital – or worse.

It was difficult to raise children because we had no money. Sometimes the mine operated only one or two days a week. Sometimes four weeks, not working, just Saturday, and you know, he'd bring in ten dollars for payday. We were in a little two-room shack in Blairmore. I never knew what I should do: pay the rent or buy the babies milk. *– Josephine Misztal*

I came from Poland. I was in the Second World War and after the war, in 1946, I came to Halifax. I came to Canada because I like the country – it's a big country. I had nobody here but I knew about Canada from school.

My friend writes me and says, "If you want a job, come over." And when I came over, I was coming temporarily – that's all. Twenty-four years later, I'm on the pension.

When I came in 1946, I worked on a farm near Lethbridge; then, in 1948, I started to work in the McGillivray Creek Mine. I worked 24 years in the mine.

I was 29 years old when I started driving the dinky at the McGillivray Mine. I worked as a contract miner for many years. After the McGillivray closed, we opened the Vicary Mine and I lost an eye in that mine. After the accident, I got a job as compressor operator outside the mine. That was a really nice job. I had lots of time to read books and write letters.

And every Saturday afternoon, we went to the beer parlour to clean up the dust, the miner's dust. *– Bill Misztal*

CHIC ROUGHEAD

I started at the old International Mine on the coke ovens. My job was brushing doors. I had to fill the cracks with clay. Every time they cracked, you had to refill them. That was a hot job. I did different things around the ovens for a couple of years – shovelling ashes and things like that. I was 17 when I started.

I worked in the tipple on maintenance, cleaning and greasing, which was about the worst job I had because of the dust. But I'd rather have done that than go in the mine. Yet these old miners, they wouldn't come out of the mine. In the winter time, if we brought miners out of the mine to help clean snow, they'd be cursing like heck.

I don't know why I started playing tennis – it was always the big shots who were playing. As a young kid, I used to go down to the club and help roll the courts. They were shale courts so you had to water and roll them each night. I used to spend hours and hours at tennis. There was an old bachelor who lived right beside the courts and although he didn't play, he helped the junior players. He had an old Fred Perry book that showed the strokes. I remember that when I was playing tennis as a kid, we could hardly afford it. I went to Trail, B.C. for a tournament and I think my mother gave me her last ten dollars.

I used to win the southern Alberta championship and I went to the provincial finals in Calgary. After juniors, I was ranked No. 2 in the province. After the war, I played tennis in tournaments all around Europe against some pretty good players. I have a scrapbook full of memories about my tennis days. But by the time I was 45, tennis had lost its popularity in Coleman. It's years ago now that I quit playing tennis. I started golfing about 20 years ago. I've had two holes-in-one. My golf game is good one day and bad the next.

ERNIE MONTALBETTI & CORBIN PIONEER, VENUTA BARATELLI

I was born in 1909 in Corbin. My father was a coal miner in Corbin and Michel. I worked underground in Corbin, Blairmore and finally in Coleman. I had tough jobs like timber-packing in the mines.

After the mines, I worked 17 years as an electrician for the Coleman Light and Water. I would go crazy if I had to work in the mines today. — *Ernie Montalbetti*

I was born in 1893 in Italy. I lived most of my life in Corbin and Fernie. I was intimately involved in the coal industry through my husband, family and friends.

Corbin was a nice little coal mine town. My husband ran the roundhouse to turn the coal trains around. — *Venuta Baratelli*

COKE OVENS

ALBERT GOODWIN

In 1908, we came to Canada from England, in the same year I was born. We came to Bellevue via Glace Bay, Nova Scotia. I was 16 years old when I started as a driver's helper at a daily rate of $3.50. After two years, I dug coal for the next 25 years until I was put up on the hill. I put 37 years in this mine (Bellevue) here and six more at the Vicary Mine in Coleman.

I was put up on the hill here during the war. That was the most foolish move I ever made in my life. I went to night school and received my certificate for mining competency. For many reasons, that was the biggest mistake I ever made in my life. Because I had charge of the whole mine, I had to work Saturdays and Sundays. The pit bosses would call me in the middle of the night because "there was a cave-in and they couldn't start

the pumps." So I used to have to get up in the goddamn night and show them how to start the pumps! When I went to Vicary, I had to pay fifty dollars to join the goddamn union. I was lampman up there at the time. That was the only way I got a job. Six months later, they asked me to go in the mine as a fire boss so I had to get out of the goddamn union that I had just paid to get in. I guess it was a little bit difficult but jobs were hard to come by then.

My dad always said the Bellevue Mine was the best mine in North America and he worked in all kinds of mines. As a hobby in the summer, I have been working as a guide at the Leitch Collieries Provincial Historical Site.

MINE LAMPHOUSE

MINE ENTRANCE

JOE BEILAN & VICTOR BELIK

There are lots of good things about mining if you're not in hazardous conditions. Some places you'd go into, they'd crack and break and then you'd be running. After a while, of course, you'd get used to it and even know when it was going to cave. Some guys were crazy and some were cowards. The live cowards are always better off than the dead guys.

In 1941, it was hard to get a job. The war was on and there were 150 men looking for jobs at each mine. I was 16. My first mine was the Mohawk and I got a job there because my dad was one of the bosses.

Bucking coal in the mine was my first job. Then, after four months, I got promoted to horse driver, pulling cars. Then, after that, I went rope riding. Then I got my "B" miner's papers. When I got my "A" papers, I was classed as a miner. In 1970, I got my fire boss papers and was made foreman of a project at the Grande Cache Mine.

Old Tony G. taught me more about mining in one year than I learned myself in 20. He said, "You never know too much. There's always room to learn." He said, "You got to dig coal with your brains and not your hands and your back. If you got no brains, you're not going to make the money." I know when I worked with him, we made up to forty dollars a day. This was when wages were twelve dollars a day. He taught me not to be scared underground.

When the diesel trains came in, the mines shut down. And when oil and gas came, coal went to hell. This was all in the early '50s. You had to be in the mine 25 years for seniority. If you didn't have seniority, you got laid off often in those days.

I was always the first customer in the bar and the owner always used to fill me up with about four ounces. He said, The first customer always gets a free drink.   — *Victor Belik*

ITTALO RUABAN

I first worked in the Star Mine near Rosedale. I was a fire boss there. Then I came to Shaughnessy and was here 23 years. I was the last man out of the mine. After it closed, I worked in various mines: Fording River, Grande Cache and Coleman. When I was working in Grande Cache, I wore out a car, driving back and forth to Shaughnessy.

I think working in Coleman was the most enjoyable. In a lot of mines, everything got to be old stuff after awhile, but not in Coleman. The mining conditions were always changing. In some places, the coal was perpendicular. In other places, you'd run into a wall of rock.

During the summer, the Shaughnessy Mine operated not much more than a day or two a week. When the mines weren't working continuously, I had the option to do some welding in Lethbridge.

I was never unemployed – I had four jobs while I worked in the mines. I was the postmaster in Shaughnessy. I was mine rescue superintendent, mine fire boss and a welder in Lethbridge.

I quit four years ago. I was getting one hundred dollars a day. I don't know what the hell the wages are now, but I'll bet they're much higher. They're thinking of starting a new underground coal mine at Kipp. If they do, my son will be able to get a job – he's worked underground.

I remember how strong Victor Belik was. One day, I saw him carrying a 16-foot mine timber about a foot in diameter. I saw him carrying this thing and I said, "Victor, you shouldn't be carrying that by yourself." He said, "Before I can get those goddamn young kids to help, I'd sooner carry it myself." He carried it all the way in.

DON LIVINGSTONE

I was born in 1917 in the little village of Hardyville, just north of Lethbridge where the Galt No. 6 Mine was located. My dad was the general manager of the company. The first thing I can remember was my third birthday: we were having a party and my dad took us all down the mine. All I can remember is the bottom of the shaft. From then on, every kid wanted to come to my birthday parties.

My dad was a professional engineer who'd trained himself. Although he left school in grade 3, he studied and got his mine manager's papers and professional engineer's papers. He became the chief inspector of mines in Edmonton. He was superintendent of Lethbridge Collieries until 1935.

I went to university to study coal mining. I worked underground before I wrote my fire boss' papers. I went on and got my mine manager's papers, then my strip mine papers.

After the war, I came back and worked at Galt No. 8 as the fire boss. I then became the chief engineer which eventually led me to becoming the general manager for Lethbridge Collieries.

When I became the GM, we had only two mines, Galt No. 8 and the Shaughnessy Mine. These mines were controlled by Canadian Pacific. Later, I worked on the initial development of the Fording River Mine in British Columbia.

There have been a lot of coal mines around Lethbridge. The domestic coal was top coal, better than Drumheller's. Our slogan was "As good as Galt."

I've owned the Bridge Golf Course for 25 years. It's a par three course. Very pretty in the summertime when all the trees are out.

JIM WILSON

I started out as a boy miner in Scotland, where I was born in 1903. When we came to Canada in 1921, I started coal mining in the Lethbridge area. My father and brothers all worked in the mines in Lethbridge. Then I moved to Coleman where I worked as a face miner in the McGillivray Creek Coal Mine.

I also worked for a while at the Atlas Coal Mine in East Coulee in the Drumheller Valley. I came back to Coleman but unfortunately I was caught in a cave-in at McGillivray in about 1955.

I was one of many crusty old coal miners who came from Scotland to work in Canadian coal mines.

MURRY'S POOL HALL

ANDY BODNAR

worked about 36 years, all in the Canmore Mine. I was born in 1914. I've been here since I arrived from the old country – Poland. My uncle used to be a fire boss in the mine. He brought me over here when I was 14 years old.

I retired three years ago after working in the sawmill at the mine. I always worked on top. I looked after all the materials needed for the mines – like props, roof bolts and ties. I also looked after all the track on top.

I wanted to go underground but they wouldn't let me. There was more money underground. I even squawked to the union about going underground but it didn't help.

In the early days, we used to make everything for the mine, including sawing our own mine timbers. When I was working in the sawmill, we started with props 16 feet long and I would cut

them to the size the miners required. The sawmill was located alongside the river, near the mine. We also cut lumber to build and repair the houses because it was all controlled by the Canmore Mines. There were four of us working in the sawmill – plus a horse. We used a horse to pull the timber up to the mine.

Me and my son built this house 15 years ago.

I went to the doctor a couple of months ago. I told him, I'm feeling okay but I'm gaining too much weight. He told me to quit eating, drinking beer and to buy a bicycle to do lots of riding. So I bought a bicycle and do lots of riding.

I used to like having quite a few beers now and then.

VIC RIVA

We had a frost last night so I'm watering the flowers, trying to save them. We usually have a heavy frost in late August, then the weather clears up for another month.

I was born right here in Canmore in 1902 and I never moved from here. Harry Musgrove and I are the two oldest people born in Canmore. My dad came from Italy and worked down in the mine. As he got older, he came up and worked in the shop for a while. Then he became a watchman.

When I started working, I started on the outside and never did work full-time underground. I worked in the shop, mostly on welding and mechanical work. If they needed me, I used to go underground to help the mechanics.

I started work in 1918 and retired in 1968. I got a clock that was presented to me for 50 years with the company. There were four of us brothers working for the Canmore Mines. I'm the only one left now. My son and grandson worked for Canmore Mines. We had four generations of Rivas working for that mine.

At that time, Canmore was a small town and you knew everyone here. It was nice, living here. This was the only house for quite a ways. Now my son lives next door to me. I had two lots and gave him one, helped him build a house on it. Canmore is getting different. You don't know everybody anymore.

WALTER RIVA

My grandfather came over from Italy about 1900. He was still working at the mine in 1940, when I started. He had four sons and they all worked in the mines. Each brother, including my dad, had more than 40 years apiece with Canmore Mines.

I worked at the mine three different times, first in 1940, then immediately after the war, then from 1949 to 1973. The first time, I remember very clearly getting my lunch-bucket and walking down to the mine with my dad in the morning, and then him showing me where to hang my clothes and where to get a lamp. I was 18.

I enlisted in the air force and, after the war, went to the University of Alberta on the veterans' scheme. As a mining student, I worked one summer for West Canadian Collieries, first at Blairmore, then at Bellevue. When I graduated, the best job offer came from Canmore. My last year at university, I worked night shift at the Black Diamond in Cloverbar. That was on the end of a shovel. I was quite glad for that experience.

I worked 23 years continuously at Canmore Mines. During my time here, we started shipments to Japan. The first Canadian coal that went to Japan came from Canmore. We had to fight hard to develop and maintain markets. I was active in selling Canmore Mines because the existing operation could not survive financially.

When I left Canmore, I went to Denison Mines to work on the northeast British Columbia coal project. A year later, I joined Kaiser Resources and ran their operations in southeast B.C.

I enjoyed the Canadian Institute of Mining because of the mining aspect. One of the highlights of my life was to become president of the CIM in 1983/84. I've held quite a few positions, including chairman of the Coal Division.

Canmore was a wonderful place to grow up – my wife and I grew up here. There were about 1,000 people here so you knew everybody, every single family and you knew every house and who lived in it.

JENKIN EVANS

Although I have lived in Canada longer than I lived in Wales, I am a true, bloody Welshman. I was born in 1902, and went into the Welsh mines with my dad at 14. For ten years, I worked at the Baldwin's anthracite mine in Wales.

Then, in 1928, we had a long strike in England. My father-in-law moved to Canmore to get established and we followed later. But it was difficult getting a job at that point because Canmore was only working two days a week.

When I started in the mine, all you could buy were those cloth hats. They used to get rotten from the sweat. I also remember that when I started, we had candles in the mine. Then we had the gas lamps, the carbide lamps. It was tough during the Depression. There were few cars so even a trip to Calgary was a big ordeal. If it wasn't for the manager of the company store then giving you credit, half the people would have starved to death.

I remember about 1935 my daughter ran into the house and said her friend had fallen off the bank into the Bow River. I ran out and saw her struggling and dived into the river. It was winter

time and there was ice forming but I could reach and save her. The government awarded me a medal for that. It was really some day when I saved that little girl's life.

There were only about a dozen Welshmen in Canmore then true Welshmen, that is. Now we have so many different people coming into Canmore and they are not coal miners. They are a different class of people. They are not the same. You haven't got the community spirit as we had when the mine was going. There were about 300 coal miners in Canmore, and we were closer knit then. In those days if a man died in the mine, it was compulsory to go to the funeral. You could be fined by the union if you didn't go.

I only worked in two mines all my life. But even with 50 years in the mines, they never carried me out on a stretcher. I was lucky. I got a clock for 40 years of loyal service to Canmore Mines. It is a beautiful clock but what good is a clock when you are retired? You want to sleep in and not wake up.

CHARLIE HUBMAN & JOE SHANDRUK

When I started, Canmore had more Ukrainians than any other nationality. And there must have been a hundred Chinese men working at Canmore Mines. We'd all be in a line on Saturday to get our pay. A hundred of us in a line and the front guy would be talking to the last guy in line. And the end guy would be hollering something to the front guy. It was a jolly old town then.

My dad was a coal miner. He farmed around Vegreville in the summer and used to come back to Canmore in the wintertime and work in the mines. My dad started No. 1 Mine in 1908 and, when he quit underground, he went to No. 2 Mine.

I started working in the tipple in 1927 when I was 15 years old. When I was about 18, I went underground, driving horses. I did rope-riding and running hoists; I went on the coal as a contract miner and I did that for years. I pretty near made 50 years at the coal mine. In 1960, I had a couple of vertebrae broken in my neck and I was done for two years. After that,

I went back on surface in the shop, repairing mine cars until I retired. I was on the executive of the union. Once you get involved in that, you just get in deeper and deeper. I was the president of the union for 22 years. I retired from the position in 1960. It took a lot of my time; there was always somebody on the phone. There was friction all the time.

In 1938, we amalgamated our local union, the United Mine Workers of Canada, with the UMWA. When Tim Buck got out of jail, he'd come up and address the meeting in Canmore. Dr. Bethune also came here, as did Harvey Murphy.

The Canmore Union Hall was built around 1910; I had an office there and every Sunday, the executive would meet at noon with the union meeting following at 2:00 p.m. When things were rough and tough, the union hall was always filled but when they were getting good paycheques, it was empty all the time. – *Joe Shandruk*

TONY STRICKLAND, PROFESSIONAL ENGINEER; **WAYNE** HUBMAN, FIRE BOSS & **JIM** MORRIS, RETIRED COAL MINING EXECUTIVE

BILL WILSON, RETIRED COAL MARKETER; **GORDON** MATHEWS, RETIRED COAL MINING EXECUTIVE & **DICK** MARSHAL, COAL CONSULTANT

MINE SHAFT FOUNDATION

JOE ORSTZ

You know, I am feeling my 80s now but if it was really necessary, I could load coal yet. I was born in 1906 in Hungary. I come to Canada in 1928 and arrived in Lethbridge to work on a farm.

A friend of mine from the old country taught me to be a miner. He taught me so well that I could get my miner's paper before I ever went into a mine. Then my friend took me underground in a small Lethbridge mine to actually show me how to mine. I come there for one week and never do nothing, just sit in the corner. Boy, was my heart pounding and I am always looking at the roof. The first time I went in the mine, by Jesus,

I was white like the snow because I was scared that the roof was going to come down. He said "Don't be scared, just be quiet." He showed me how to load and to bore and load the holes. After that I come to Wayne, Cambria, Rose Deer and the Shaughnessy Mine and that's where I finished after 27 years in the mines.

I sure enjoyed being a coal miner. All my miner life, I liked it very much and thanks to the God, I only had one small accident. I always remember my teacher who said "Safety first. Always mind the roof. Don't worry about the coal. Whenever you lose your life, nobody is going to replace it for you."

ANDY KRALIK & **S**POUSE

I came to Canada in 1926 and first started working at a little mine in Bienfait, Saskatchewan. From there, I went to New Waterford and worked in No. 10 and No. 16 Collieries. I didn't stay in Nova Scotia because there wasn't much work and the pay was poor.

In 1928, I came here to Drumheller. I worked right through until 1955. I quit then because I wanted to operate the farm I bought in 1943. I could walk from my farm to the Red Deer Valley Mine. The mine workings went right under my farm. They even had an airshaft on my land and sometimes put materials down for the miners.

Oh yeah, it's much better to work on a farm than be a coal miner. You get good air, you know. My son now looks after my farm. He has a farm right next to mine. In 1943, a farm was not too high of a price.

We have been married 63 years. I was married in the old country, Czechoslovakia, in 1923. I worked in coal mines in Hungary and Czechoslovakia. Oh, you betcha your life, I was an experienced miner.

JOHN YORKE

I was born in 1905 in Lethbridge. I started mining in the Galt No. 6 mine in Lethbridge when I was 16 years old. My father, who was born in Czechoslovakia, was a coal miner in various mines in the Crowsnest Pass and Lethbridge areas.

I worked in the Drumheller and East Coulee mines but also worked in the Lethbridge-Shaughnessy areas as well as in mines in Coleman and Medicine Hat.

I held a variety of jobs from trapper, horse driver, rope-rider, mining machine operator, fire boss and lastly as mine tour guide at the Atlas mine in East Coulee. Foremost, I considered myself an expert at breaking-in and driving horses. Once, I drove a three-horse spike team hauling coal underground.

During the summers, when the mines did not work, I played professional baseball for the Galt Miners in Lethbridge and for a team in Vancouver.

I wanted to quit the mines when I was 70 but they asked me to do tours at the Atlas. I think it was time to quit at the age of 75 after 58 years in the mines. In the summer time, you get outside. When that old winter comes and it starts to get cold, you head down that hole and you get used to it. I enjoyed the mines but there were some tough times.

ANGELO CLOZZA

RETIRED COAL MINER

SID McMULLEN

I was born in 1914 in Edmonton. My destiny was to be in the coal business when I was adopted by S.L. McMullen and his wife. My stepfather developed the Midland Mine No. 1 in 1912.

S.L. McMullen had a bum heart and had to live in Victoria most of the time because of the lower altitude. I came to the mine after a couple of years in engineering at a B.C. university. I found I could learn mining engineering at the mine – practically instead of theoretically. I decided to work in various jobs in the mine and learn mining that way.

I first went underground the summer of 1934. I did track laying, timbering, you name it. Oh yeah, I shovelled a lot of coal.

I had my fire boss papers and was going to write for the pit boss when I joined the army. S.L. McMullen died in 1941 while I was overseas. I had just signed up for the duration so it was 1945 before I could return to the Midland Mine. I met my wife in Holland during the war.

After her husband's death, Mrs. McMullen became president of the company. She had a lot of advisers and staff but she didn't get good co-operation. I think a lot of people were just robbing the place. When I got back, I became vice president and general manager. I guess I was successful because I kept the mine running until 1959. The market for Midland Coal was good until 1947 when oil was discovered. Midland coal was domestic coal, used for household heating and small buildings – there were no big contracts. For most of the time I was general manager, we were faced with declining markets. I toured the country, trying to develop new markets.

Before I left Drumheller in 1963 and after we closed down the plant, I got the bright idea to make a park out of company land. This would provide a place for kids both young and old to run around. I offered 1,700 acres of land to the provincial government with very few strings attached.

ANN AND BILL HAYDAMACK

I never worked too much in the summer because I was golfing. I started to golf in the early 1920s. In Wayne, there were quite a few golfers and we had a Golf Club. We used to play for the Shield. I was the champion for several years.

I worked on building the first road to Drumheller. Some fellows said to me, "Why don't you go in the mine? It's nice and warm in there." In 1919, during the last part of May, we had such a blizzard that you'd think it was the middle of January. You never saw anything like it: cattle were frozen, trains went off the track. That's how I got in the mines.

I was 19 years old when I started at the Jewel Mine in Wayne. I thought to myself, I want to make a career out of this mining business – I like it. I really like machinery. After Wayne, I went to East Coulee in 1928. I finished up in East Coulee in the mine about 1973 or '74.

In the mine, I worked up my reputation: a qualified miner, a qualified mechanic. When you have a bad relationship with your workers, you're not going to get on any place. I got on because I treated everyone the same. Except that anyone who worked for me had to know how to sing. – *Bill Haydamack*

FRANK ZAPUTIL

was born in 1908 in a little coal mining town in eastern Washington State. My dad was looking for a new mine to work in and we ended up in Wayne in 1920. From that time to now I have been in the Valley. In 1920, that's when I had my first fight, too. I wasn't halfway down the railway tracks with the suitcases and some kid challenged me and he got it.

Oh yes, I went to school in Wayne until I was 15 then I hit the pits, the Jewel Collieries. I was a motorman for 20 years on the main haulage. My dad was responsible for me going into the mines because I wouldn't go to school. So he said, "Okay, I'll fix that." Well, I wasn't sorry about it because I wasn't satisfied with the teaching at the school. They were 17th-century teachers. They were still wearing the long skirts and celluloid collars.

In 1938 the Jewel Mine closed in Wayne and reopened in Cambria as the Western Gem & Jewel Collieries. When I moved here, I studied up and made my third-class fire boss certificate.

I closed the mine here in Cambria in 1950. I was the boss on the job. When everything was taken out from inside, I closed-up her up the airway and the main entrance. After Cambria I fire bossed for another 10 years in the Atena and Murray mines. When the Murray closed in 1960, I came home and I told the wife, I said, "By golly they are all closing up everywhere I go." I said, "Maybe I should stop going anywhere." I tried another mine, the Highgrade in Drumheller and it quickly closed so I said, "That's it."

I have always lived in this house since I came here in '38. We moved this house from Wayne and it only cost me sixty-five dollars. Originally it was only a two-room house. Cambria was filled with people when the mine was operating. We had a hotel, post office, two grocery stores and a bus stop. When the buses went out any day, they were full. The hotel was the only place you could get beer when it was rationed. There was more beer there than you could throw a stick at.

ANGELO BERLANDO

O h yes, I was brought up an Italian boy. There was nobody like Angelo to my mother. She used to chase the girls away. I remember that. "Leave my Angelo alone!" That's the way Italian women are. I was her favourite son and she didn't want to lose me.

I was born in this house in 1926. My dad came from Italy in 1904. When he first came, he opened up the Moonlight Mine about one mile from Rosedale. He worked mainly in the Star Mine. He died when I was 10 years old.

I played baseball, hockey and I boxed. Sports were the biggest thing in my life. If we hadn't played sports as kids, we would never have gone anywhere. Our Rosedale ball team was Alberta junior champion for 10 years. And my brother and I played for about 10 years in the Drumheller Miners' hockey team. That was our life. We used to work in the mine, come home, wash up and head for the ball field.

I started in the Star Mine in 1942, picking bone. Then I went underground where I pushed boxes and drove a horse. Then I came outside. Most of my years in the coal mine were outside, loading box cars. I was at the Star Mine from 1942 to 1957, when it shut down. Then I was offered another job at the Sheerness Mine, but I went road-building for the next 30 years.

Aerial was the name of this mining camp because the Aerial took the waste rock and coal from the tipple and dumped it in the coulee. At one time, the men used to ride in buckets across the river. Then they built a narrow suspension bridge for the men to walk across. When it was windy, the bridge used to whip up and down. It was no fun crossing on that bridge at first. There was nothing to hold onto. I wouldn't have traded coal mining for the world. I loved it there, I'll tell you. You know, what I liked about the Star coal was that it really shone.

JOHN GRAHAM

I was born in 1912 in Calgary. My father moved us to Drumheller in 1915, when the new coal mining town was starting up. He trailed 85 dairy cows to Drumheller, and he milked them twice a day on the road. My mother had us four kids in a covered wagon. It took us two weeks from Calgary to Drumheller.

Dad used to deal in horses and he started raising ponies for the coal mines. That's what we made our living at, you see. The mines would want ponies the height of their coal seams. In East Coulee, most of the coal seams were 50 to 54 inches thick. We sold ponies to all the mines in East Coulee, some to Drumheller, some to Wayne.

In later years, I used to go in the mines and help them break the ponies after we'd sold them. I often broke horses in at the Atlas Coal Mine. After the coal mines finally were mechanized, we started raising horses for the rodeos. The mines here stopped using horses in the late '40s and early '50s.

I was tempted to be a driver in a coal mine quite a few times but there always seemed like there was too much to do at home. I was often tempted because of the pay. That's one thing about mining: in the tough times, you didn't get much work but when you did work, it was one of the best-paying jobs around.

BEN LOWTHER

I have to put it this way: I liked farming best although I do like mining. You know, I can't knock mining. I have to look at it from the Depression and right up to today. I was born in 1900 in England. In the fall of 1929, I started to work in East Coulee and I've been there ever since. I first worked for my wife's dad who needed some miners.

During the Depression, I could go into the mine and knock out between seven and ten dollars a day which was enough to keep us going for a couple of weeks. We could make enough in the winter to keep us going through the summer, whereas the rest of the people in the country were really having a rough time.

I had 46 years of coal mining including the following jobs: contract miner, building the tipple, duckbill loader operator, fire boss and mine manager.

In the early days, the mines produced lump coal. There was no market for the small coal. The lump coal market was the big thing so the mines all tried to get a higher percentage of lumps. You know, you put in a couple of lumps and they would get a little bed on the bottom and then the fire would last until morning in the cold weather. Coal is a good heat. It is a very pleasant heat, either in a furnace or a potbelly stove.

The coal industry in the Drumheller Valley never did seem to slack off – it just seemed to fade away. As the mines were closing up, miners were retiring and there were no new miners coming in. Coal mining in the Valley just seemed to pass away.

JOE LASLOP

I was raised in Czechoslovakia. When I came here in 1948, I started in a small coal mine up Willow Creek, not far from here. My first paycheque bounced. That was quite a surprise to me.

In 1948, I got a job with the Red Deer Coal Mine in Nacmine. I ended up at the Midland Mine with Sid McMullen. When the Midland and the Crown amalgamated, I ended up with the Crown Mine. Then the Atlas and Crown joined together. That's why I'm here at the Atlas.

I didn't go underground because being a mechanic is in my blood. Being a welder, mechanic and shopman ran in my family. My only brother was also into the mechanical end.

I worked for the coal company all my life, but always on top. You didn't get the excitement and happening that you got underground. In the old days, you had to buy your own tools. Your payroll number was your tool number. When a miner had to have a new pick handle, he had to pay for it. He also had to pay to sharpen his pick.

Around here, the only thing that grows on the coal is the tumbleweed. There are lots of old-time miners who worked in the Atlas and the Murray. They're all on pensions and scattered all over Alberta.

TIPPLE BUILDINGS

OMER PATRICK

CARL LAMSON & **LEON** MEHL

REX CHRISTENSEN

TIPPLE FOREMAN – MONTGOMERY MINE

VINCE MILLER

CLINT MILLER, DRAGLINE OILER & **DAVID** BOEHIKE, DRAGLINE OPERATOR

JAMES NORDBY

We leased this mine in 1956 because my friend came to me and said "Let's give coal mining a try." My wife and I moved out here from Ryley. Then in 1962 we bought the mine. It's always been called Dodds Coal Mine. It is one of the oldest mines in this district of Alberta. The mine started in 1905 as an underground operation. Once there were 62 men working here. When we first started, the Black Nugget Mine was operating nearby. There was another mine nearby where they stripped with mules in the '20s.

Our coal seam varies in thickness from 5.5 to 7 feet and is between 16 and 26 feet deep. We sell mainly nut and pea coal. There is not a large market for stoker anymore. It's good coal with an ash between 6 and 8 percent. Our sales averaged between 15,000 and 25,000 tonnes a year. Black Nugget, the largest mine, sold anywhere between 40,000 and 60,000 tonnes a year. I operated this mine for about 35 years.

I wouldn't have stayed here if I didn't enjoy mining coal. In fact I did better at mining than I did at farming. It gets in your blood, you know. I was my own boss I liked doing the stripping and running cats. Now, a contractor is used to do the stripping. I don't have any more to do with it now. I have leased the mine out to my two sons-in-law. But I pester them and I still give them a little advice on the stripping. If you start moving dirt twice, it can cost money. When we first came here, coal was worth only $2.50 a tonne. The same coal today is worth twenty-two dollars a tonne. The trade name for this coal is Red Flame.

RUSS BISH

HAROLD ROSS

KURT McMAHON, SUMMER STUDENT; **WAYNE** McMAHON, RETIRED COAL MINER
& **JERRY** McMAHON, WELDER

PAINTEARTH MINE

TIM JENISH, MINE MANAGER; **TIM** NEILSON, OILER

& **BRUCE** JACKSON, DRAGLINE OPERATOR

BUD MacFARLAND & JOHN WAGNER

MINE BUILDINGS

ART VISOTTO

ERNIE MORLEY

DUNCAN BROWN

I was born in 1903 in Newcastle upon Tyne. I studied mining engineering but there were no jobs for engineers so I immigrated to Bienfait, Saskatchewan in 1928. From there, I moved to Alberta, Mountain Park where I worked as a miner, ventilation man and a fire boss. After eight years, I got kicked out for trying to form a union for fire bosses. In the early '40s I became a pit boss at the Saunders Creek Mine in the Nordegg area. It was the only time the mine made money. I also worked at Nordegg and Alexo mines.

When I left Alexo, I became the mines inspector in the Drumheller area. There were 57 operating mines that I had to inspect including 21 in the Drumheller Valley. My job was to check every mines' working places and the ventilation, according to the Coal Mines Act.

When I was inspecting the big mines in the Pass, I would go underground in the morning with the miners and the mine manager. Between the manager and me, we would have only an orange or a candy bar to last us all day in the mine. We would walk continuously all day throughout the dirtiest parts of the mine. I would do the same the next day at another mine. My biggest concern was gas.

I was 16 years in the Pass and Lethbridge areas. Then I was transferred to Edmonton and eventually became director of mines before retiring. I got 51 years of mining. After retiring, I did some prospecting for coal but my hobbies are painting pictures and feeding the blue jays everyday.

When I first came to Alberta, the mining industry was all coal men, now there are no coal boys left. Running a mine had its peculiarities. The problem was that the miners never looked upon changes as beneficial. All coal miners had an innate suspicion of the mine manager.

HOMER BISH

My father came to Canada from Oregon and homesteaded on the Battle River near Forestburg, where I was born. In the early '20s, he started mining along the river. Eventually, my brothers and I began to work with him and later bought a half-interest in the mine. In 1949, we sold it, just after converting it from an underground to a strip mine.

I stayed on at that mine. I had an underground mining ticket but no strip-mining ticket. Eventually, I obtained a strip foreman's certificate, then a strip mine manager's ticket. I became a mine manager under an American who was there to develop the mine and teach us new strip-mining methods.

In 1957, I helped Alberta Coal Company start up Utility Coals. I stayed there until 1962 when I transferred to Wabamun. I managed the Whitewood Mine until 1974 when Saskatchewan Power was expanding its use of lignite coal for thermal power generation. Eventually, I became the manager of mining for the Saskatchewan Power Corporation in charge of all their mining operations. I stayed there until I retired in 1982. I enjoyed the people and the challenge of mining.

Coal mining – especially prairie coal mining – was affected directly by the economics of the country at the time. When the price per ton of coal was low, the difficulties of operating and the small margin of profit were reflected on the equipment used and the number of men hired. It's an exercise in economics.

It's strange: I've had people work for me with only grade 4 or 5 education. If they'd had access to university education, they would have been very, very clever people. I had a welder who worked for me: if you told him to build something, he immediately knew the size of iron, the type of welds – fantastic the type of people you'd pick up. Electricians – I had some fantastic electricians. When you get into bigger equipment, they're pretty important people in a coal mine.

PETE PAQUETTE

O h yeah, he's quite the goose. And he can get pretty mean. He chases the dogs. Real mean.

I was born on a little farm near Jackfish Lake in 1924. I was 16 years old and living at home when my dad got sick. He was working at the Victory Mine at the time. I went in his place and worked about three weeks until my dad got better. A year later, they called me back to work. They wouldn't let me work underground so I quit and went to Nordegg Mine to work underground. I didn't stay long because I got homesick.

I came back here, got on the underground and worked continuously until 1961. The Whitewood Mine started up in 1962. I got on here through my old manager.

I've got about 20 years underground in various mines around Wabamun, the Coal Branch, Edmonton and Drumheller. I worked about 25 years on surface at the Whitewood Mine.

On surface, I started on the pumps, taking water out of the coal pit. Then I got on a coal hauler for about three years. It looked easy but it wasn't. I finally got on the draglines as an operator. I really liked the old friction-operated draglines. Now I operate a coal shovel and drive a truck when need be.

I want to work as long as I'm able. When I retire, I'll work on my acreage where I have some horses and chickens.

JOE JOHNSON

I came to Canada in 1929. A friend looked after me and got me my first job on a farm north of Edmonton. I started on a mink ranch and then came to Wabamun in 1946. Things were pretty tough. The first time I worked in a mine was in 1949.

Then I worked cutting ice for the CNR between working for the mines. I worked at the Victory Coal Mine a good many years but never underground. My first job was shovelling coal into box cars. Then they put me on to loading coal. In 1962, I started at Whitewood Coal Mine. First I worked on the pumps day and night, including picking rocks. Then I went across the lake to the Highvale Mine as an oiler and became an operator. I like the excitement of working on the tipple. It was going strong at the time. We were producing all sizes — lump, stoker and slack.

You can only do so much when you are 75 years old. I have an accordion and can play. When I was younger I had a button accordion. As the years went by, I got a piano accordion. When I was younger, I played for the odd party.

RUBY KOFLIK

I've lived in Wabamun all my life, except for the time I went to business college in Edmonton. My father was a blacksmith at the Victory Coal Mine. I started with the Whitewood Coal Mine in July 1962 and so far, I've spent 25 years in the mine office. That's when this mine first started. I'm still doing the same work but now we have computers to help look after the payroll.

I was with Manalta Coal from 1962 to the end of 1986, then Fording Coal took over the operation of the Whitewood Mine.

We've had a lot of fun over the years. I know most of the people who work at the mine but not all of them. A fellow came into the office recently; he said he'd been with the mine ten years, but I'd never met him before.

I used to know a lot about coal mining but it's become much more technical and I'm no longer familiar with the terms.

DON KINGDON & LOUIE CARRIERE

I was born in 1930 in the mining camp at Bienfait, Saskatchewan. My dad worked at the Bienfait Mine. I got started in coal mining when my dad had his own small underground mine right within the town limits of Bienfait. I ended up as a manager with Alberta Coal (Manalta Coal) at the Whitewood and Highvale mines in the Wabamun areas.

When I first quit coal mining, I found it pretty hard. You go to work everyday for about 25 years and never miss a day for hardly any reason at all and then suddenly 'bango' the load is gone. Your engine races away from you. So now I raise cattle and it is a pretty good way of life. The best side of coal mining was the people and their lighter side – the different capers and funny things that would happen with a bunch of men around. There is also an element of glamour in coal mining. – *Don Kingdon*

I was born in Legal, Alberta, in 1926. I was in the coal mines all my life, forty-some years. In 1944 when I was 19, I started stripping coal at Cadomin with Mannix Construction as a cat operator and ended as a shovel operator within six months. I then moved to Forestburg. They put me operating the Page 8 yard dragline, then the biggest machine in the country. After that, I worked in the Estevan and Wabamun areas for Manalta Coal.

Equipment always used to fascinate me. Even as mine manager, I used to go out to the pit every day, load a little coal and run the shovels and draglines. I really enjoyed that. I think the people I worked with appreciated it when they saw the mine manager could run the equipment. They'd say, "Hell, you can't do this, you can't do that." I'd say "Get off the seat and I'll show you!"

I got my own airplane, a Cessna 172. I still play hockey. Lately, I played in California with the Edmonton Old-Timers, all old pro hockey players. There were 38 teams and we ended second best. Oh, I always played hockey. – *Louie Carriere*

LYLE HOBBS, MINING ENGINEER & **DAVID** DENTON, SHOVEL OPERATOR

VERN TARNOWSKI

RECLAMATION SUPERVISOR – HIGHVALE COAL MINE

JIM FERRIER

RANDY DAVIS

ART McCLURE, PRODUCTION FOREMAN; **RAY** LeGROW, MINE SUPERINTENDENT

& **DENIS** GASPÉ, MINE MANAGER

CHARLES OSTERTAG

I was born in 1910 in Alberta. When I was 15 years old, my dad, he got hurt in the mine so I had to get out and make a living for the family.

I started in the Evansburg Mine picking rocks on the picking table. Then I helped lay track underground. From there, I was a timberman's helper with my dad. Then I got my papers and went digging coal. Evansburg Mine shut down in 1936. Then I moved to the Coal Branch and worked on the Mountain Park tipple. I felt right at home working underground. Although I wasn't a very big guy, I was really tough — tough as long as you last! I had my own mine, the Pembina Peerless Coal Company down the Pembina River. I started it all on my own. With the little money we saved in the Coal Branch we came to Evansburg. I sunk all our money in here and I had to work harder than ever.

I did surface mining for 28 years until 1973. I am not going to dig any more coal at my age. I have had enough of the coal! Coal don't look black to me anymore, it looks kind of brown.

WES BREMNER

ALEX MATHESON

RETIRED COAL MINER

KEITH BOUDREAU, SHOVEL OPERATOR & **DON** WALLACE, TRUCK DRIVER

HARRY DIRK & DALTON ROY

GREG LaVALLEE

KYLE McCRACKEN, DOZER OPERATOR & **DALE** GRAVEL, SERVICEMAN

CHRIS RACE, HEAVY DUTY APPRENTICE; **DON** MOOREHEAD, WELDER; **DARYL** DAVIS, MECHANIC; **GORDON** BANCROFT, MECHANIC & **KIRK** MORK, MECHANIC

ANNE MARIE TOUTANT

WILL MORTON & **RAY** MARTIN

DAVE SCHWARTZ, **BILL** LEE & **ADAM** JEBALLA

LLOYD LAYES, **BOBBY** STEWART & **MALCOLM** McNEIL

SHANNON PHILLIPS

BILL MILLWARD

GEOLOGIST — SMOKY RIVER COAL MINE

LOUIS DOUZIECH

COAL MINERS

OF SASKATCHEWAN

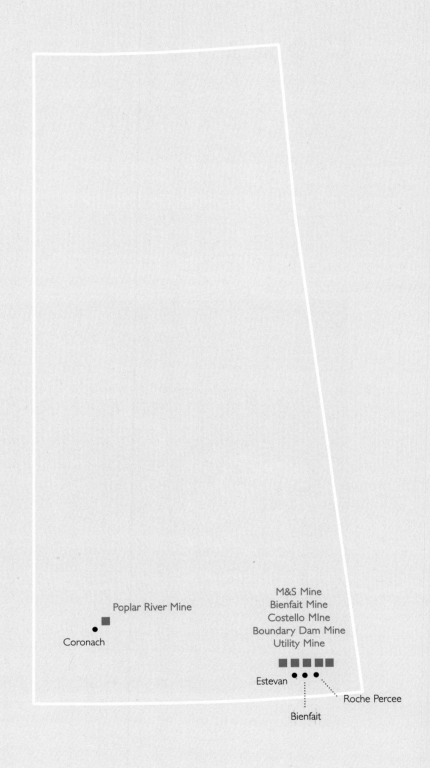

M&S Mine
Bienfait Mine
Costello MIne
Boundary Dam Mine
Utility Mine

Poplar River Mine

Coronach

Estevan

Bienfait

Roche Percee

● PHOTOGRAPH LOCATION   ■ COAL MINE LOCATION

GERRY BEAUBIEN, OILER; WAYNE KELLEY, MINE MANAGER
& WES HEATCOAT, DRAGLINE OPERATOR

I t was quite accidental that I got into coal mining, but I haven't regretted it. It's been good to me. I like the size of the equipment. That's one of the things that attracted me to this place.

I was born in Grandfalls, Newfoundland in 1948. I started with the Mannix company in 1968, in Newfoundland. In 1970, when I went to Estevan as a mining engineer, I started at the Costello Mine. Then I floated between the Costello and Utility Mines for a few years. In 1976, I was made manager at the Utility Mine.

When this project started up, I was put on one of the bid teams to work on the mine. I spent 14 years in Estevan and then, in 1984, I moved to Coronach. After signing a contract, I pushed my way into a permanent job. These days, I deal with people and paper which wasn't why I got into the business.

I think we all prefer to go back to where we first started out, especially when you get into a fight with the union or over a grievance. In engineering, you don't have to deal with that – just pass it on to the mine manager. But mostly, I enjoy this too well to ever give it up. If I had to go back to a lesser position, I don't think I would be too happy.

At Poplar River Mine, we're mining 3.6 million tonnes of coal a year and moving about 29 million bank cubic metres of overburden to get that coal. About 800,000 tonnes are taken off with scrapers and the rest with the two 90-yard draglines. The coal is loaded with the 195 BE Shovel. Six trucks – 150-tonne capacity trucks – haul the coal to the tipple. We have 140 union people and 21 managers to operate the mine. – *Wayne Kelley*

DON WILSON

EMERSON JONES

HARVEY BOLES

I was born and raised here. My dad worked in the coal mines during the early years at Roche Percee. He was blinded in the right eye from picking coal.

I trucked coal when I was 15. In '39, everybody went off to the war. I hauled coal all around this area within a sixty, seventy-mile radius on the Canadian side of the border. Every farm, every house, every building in town burned coal in those days. We trucked coal and shovelled it down all the coal chutes into the cellars and granaries, wherever they stored it.

I lied about my age and got into the army when I was 17. After the war, I worked at small mines. In 1950, I went to Roche Percee and we set up the 350 Marion stripping shovel there. I worked there until 1960 when we moved on to new reserves at Costello. Then, in 1964, the Mannix family bought the Costello and I spent the rest of my time here.

I think the interesting part is the equipment, the iron. Yeah, to me it's the iron. I sure like equipment.

The old 5W, boy, I could tell you stories. By Christ, that was a terrible machine. Friday night would come and you could bet your bottom dollar you'd be halfway home from the mine and they'd call you back on the two-way radio, every time. It never failed. That damn old 5W. It's now dedicated to a museum, if they ever get one going.

After working on equipment, I went into the warehouse. Then I was purchasing agent. I came back as a pit foreman and retired as the mine manager. I liked working as pit foreman the best. You're out on the spread, you're out where the action is. You don't have managerial or supervisory problems. Yeah, I enjoyed it. That was a good job.

JOHNNY WETSCH, RETIRED SHOVEL OPERATOR & CASPER FIEST, RETIRED WELDER

I was born in 1909 in Cannonball, North Dakota. In 1930, I was working on a farm and got laid off. I didn't have any place to go so I came to work at the Truax-Traer Mine. Most of the time, I worked as a shovel operator. I had other jobs too, such as oiler on the shovels. I worked at various surface mines in the Estevan area and did some stripping at the Sheerness Mine in Alberta.

I didn't like it all the time but I liked it most of the time. It was a steady job for raising a family.

I retired in 1966, when I was 62 years old. — *Johnny Wetsch*

Since I retired, my hobby is building pumpjacks and windmills out of scrap metal. I betcha I made 20 of those things... I gave them all away. All the kids got some and I sold the odd one.

I started at the Truax-Traer Mine near Estevan the winter of 1938. That mine was one of the first surface mines in the area. My first job was laying railway track for haulage motors.

Then I was moved to the shop area where I learned to be a welder. I was a welder for 36 years. I liked welding because there were always different things to do and we were always inventing stuff. Most of the time at the mines, I was on bucket maintenance, repairing the breakdowns on the buckets and doing general maintenance work at the pit.

I'd go back and do the same thing all over again. There was something different every day. You could always learn something. — *Casper Fiest*

JOE HIRSCH

N ow that I am retired, I don't do much of anything. I go downtown to meet the fellows every day at the shopping mall. I was born in 1907 in North Dakota and my folks moved up here in 1909. My dad worked on the railway.

In 1930, I started out in underground mining. In 1931, during that strike, I worked at Eastern Collieries. I had all the jobs underground. Underground mining was all right because, at that time, you couldn't get any other work. We worked hard but had lots of fun.

In 1960, I helped build the tipple at the Costello Mine. From 1960 until I retired in 1979, I was tipple foreman, looking after crushing and screening at the tipple.

The tipple is a crushing and screening plant. You can make the coal any size you want. There are about eight vibrating screens. After the coal was sized, it went into rail cars for delivery to the markets.

I was great for production – I was always pushing her along. At the tipple, we loaded the boxcars. It was dusty in the tipple but I got used to it – I liked it very much. I didn't like it all the time but I liked it most of the time. It was a steady job for raising a family. It provided a living but not a good living.

JACK HILL

I was born at Oxbow, 40 miles east of Estevan in 1916. I'm seventy one years old. I worked in the coal business about 40 years. I started out in 1939 at the Old Western Dominion office on payroll. My father was a coal salesman for the Great West Coal Company and he got me a job in the office. He was in business for himself, handling grain and selling coal. In those days, you did a little bit of everything. You would do a little bit of bookwork then you shovelled a little coal and weighed some trucks. You had to be diversified.

At the Old Western Dominion, we lived at the mine site. There were 55 houses there and they had a big hall. There was something on at the hall every night. You had a pretty heavy smattering of Ukrainians and Polish miners and a preponderance of people from Lancashire, an English mining area. I found them to be good people but I found they had a quaint little way of running their own show.

In 1960, I went to Costello Mine and worked there until 1978 when I took early retirement. The 40 years slipped by pretty quick. I was sort of the manager's right-hand man. I liked to call myself paymaster or even water boy – whatever. Primarily, I was just payroll.

I never had any trouble with the men. I liked the miners though some were bellyachers. On payday, they would claim they were a day short. I could tell you who was going to say that. I outfoxed a lot of guys but it took a lot of baloney to do it. By and large, I always got along. You could get along with anybody if you put your mind to it. Now that I am retired, I do as little as possible – watch soap operas and fiddle a bit in my yard although it don't look too bright.

MILLARD HOLMGREN

Big equipment gets in your blood. It's a little different now but in my day, there was no way you could go to school to learn to look after big equipment. You learned through experience.

I was born right here in Estevan in 1915. It was the mechanical end of mining that was the first drawing card for me. They built the first strip mine in 1930 and I was going to collegiate then. I went down every night to watch them erect the equipment. I was there on a Saturday when they took their first bucket of dirt to fill in the ditch. I really thought that was something. I spent a lot of time, from then on, down at the mine.

Truax used to deal at my dad's garage, so I got to know him. He needed a mechanic and that's how I got my job. I started at the Truax-Traer Mine in 1935.

Then I went to shovel operating. From there, I went to pit foreman, then to assistant superintendent to superintendent and then back to mechanic again. I finished as master mechanic at the Utility Mine. I was responsible for moving the 7800 Dragline from the Costello over to the Utility Mine. The next spring, they dropped the boom on the machine and smashed it all to hell. They bought a new boom and I gave them a hand putting it on.

In moving the dragline, we only went about seven or eight miles, but we crossed two highways, we crossed two railways and we crossed the river and then we crossed the ravine out at the power plant. It was as simple as it sounds.

At one time, I supervised 300 men. I don't think people today realize what it was like to work with men in those days. Some men you led, some men you drove and some men you coaxed. Today, it's all, you might say, in black and white. In the early days, you played it by ear. Dealing with people was the toughest job I ever did. You can fix machinery but you can't fix men.

LAWRENCE STEPHANY

DRAGLINE PIT

JOHN LOCHBAUM

Bienfait is bigger now than it ever was. There used to be only three or four stores here, but they're not even rough anymore. Years ago when there were 400 single men here, it was rough and there was lots of gambling.

I was born in Russia in 1918. I've been hanging around the coal mines since I was a young boy – 16 years old. Oh, I'll tell you, I started in the Shand Mine in the fall of '34. I was only 16 years old but I worked on top. In 1939, I got into the M&S Mine as a driver and a timberman. Then I went into the army until 1945.

When I came back, I worked until the underground closed. As a matter of fact, I'm an old dragline man now. I've run dragline for 33 years since the M&S Mine started stripping. I am still there and I've got only one more year before I retire. I've seen enough of the mines but I don't know what I'm going to do when I retire.

As far as the old dragline goes, I guess I'm the only cowboy left. The rest of the operators are all younger now. I broke in a lot of men – oh gosh, I don't know how many. You get used to it, just like driving a car. Some pick it up quicker than others and some don't give a hoot, if you know what I mean.

I've pretty well worked steady because I'm on the dragline, but there were slack periods in the summer months. I wouldn't say operating a dragline is hard work. If your eyes don't see it, your ears pick it up. I think it's just my line of work. Sometimes it can get a little monotonous, especially during the midnight shift.

We always had a garden. You had to with five children.

JIM SAXON

BRYAN DAAE, TRAINING COORDINATOR & **JOANNE** BROKENSHiRE, DOZER OPERATOR

BOUNDARY DAM MINE

LANCE MARCOTTE

KELLY AVERY

MIKE MATIJEVICH

I came to this country in 1929. I came to Bienfait in 1932 from Quebec, heading to the Copper Mountain Mine in British Columbia but I ended up on a farm, stooking. I met a young guy from Brandon who recognized my clothes from the old country. He told me I could find work in the coal mines at Bienfait. I went from mine to mine and finally got a job in a gopher hole.

In 1932, I landed a job with the M&S mine. The first year in the mine, I got hurt trying to help a guy put a rail car back on the track. When I got hurt, the doctor said, "How do you feel, Mike?" I said, "I feel pretty bad. Give me poison. I can't stand it."

I understand everything what's in the mine – what are the rules for machines and electricity and every goddamn thing.

I ended up working 36 years for M&S Coal Company. I eventually became the night watchman. When I retired, I just stayed here in Bienfait. I moved into the hotel and have lived here ever since.

Oh yeah, there used to be lots of gambling here in Bienfait. I remember one time over at Alex Ronyx's place, we started playing poker. He cleaned me out all the time, that son-of-a-bitch. I lost seven hundred and fifty dollars and I had to give him a cheque.

METRO KATRUSIK

I wasn't quite 15 when I started — that was in 1932. I got a job in a gopher hole along the river, pulling coal with a horse at seventeen-and-a-half cents an hour. They treated people worse than horses. If a horse got killed and it was the driver's fault, the driver would be fired right now. But if a driver got killed, so what? There were always about a hundred men outside waiting to take his job. It didn't cost them a bloody nickel. But a horse? You had to go out and buy a bloody horse.

On surface, I started out at the tipple, running a loader to load boxcars. In 1950, I went to work on a new dragline. Then, in 1960, I went over to the Utility Mine and I was there until 1982 — 22 years.

My dad had wanted me to get an education, but times got tough. He didn't have an education himself — he could sign his name but that was about it. I myself was pretty good in school but then the '30s came along and I figured, What the hell was the use? There were teachers working in the coal mines. My dad was mad as hell when I quit school, so I used the argument that even

teachers were glad to get a job in the coal mines. He said, "Look Metro, times aren't always going to be like this. When times change for the better, them teachers ain't always gonna be there but you're going to be stuck." And that's exactly what happened.

I got to like mining, especially once I got involved with draglines. I like machinery. I know when I ran draglines on the nightshift, a lot of guys would curse me. I was in my glory — I'd sit there all night and not even yawn. When I worked nightshift, I never came home to jump into bed — I used to build. I built all my own houses. I built three and one for the boy. I built the first one in 1950.

I have one son who's shift foreman with Manalta. I've been foreman for the last six years. One time I was assistant manager, until I got sick. I was on the Bienfait town council for many years, then I was mayor of Bienfait for six years. I retired from the mayor's position four years ago.

We came here from Poland in 1910, when I was four years old. My old man worked at the M&S mine and we lived at the mine camp. I started at the old M&S in 1921. I was only about 14 years old and I got a job trapping – opening the doors for the horse drivers. I worked at that for about six months, then I went driving a horse. Then I came up here and got a job at a small mine near Bienfait called the Bienfait Coal Company.

In 1935, I bought the pool hall in Bienfait. I picked up pool tables in North Dakota and fixed them up. And I started cutting hair in the pool hall. I got my barber's license in Regina and I'd practise on farmers. I got so I could cut a head of hair in 10 minutes.

In 1950, I sold the pool hall and worked on the tunnels in Kimano, British Columbia. There were 5,000 men working on that project. In my spare time, I cut hair.

After Kimano, I got back into coal mining. I worked at the surface mine at Taber, Alberta and then went to Hillcrest where we spent the winter. My wife didn't like it so we came back to Bienfait and I bought the pool hall again.

By cutting hair, I got in very good with the miners. I made a lot of money off them from cutting hair and playing poker. Lots of times, I would have four poker games going on in the pool hall. I was raking off 25 cents from each pot and that sometimes amounted to almost ten dollars an hour.

Bienfait used to be a gambling, booze-running town. I'll tell you the honest-to-god truth: I used to be about the best pool player around. I used to take on anybody. I would beat everybody in Estevan. I can't play anymore since I had an operation on my eyes for cataracts. Oh hell, I used to play guys for fifty dollars a game.

Some awful queer guys worked in the coal mines. One time, I got a guy in here: he'd been in a German concentration camp. The Germans had beat him over the head and he was a bit off. I used to give him a nice brushcut. This one day, he gets out of the chair, looks in the mirror and says, son-of-a-bitch, that's an awful fine haircut! He gets back in the chair and says, let's have another.

Oh Christ, my wife left in 1967. We had four kids.

I still cut hair – I got a barber chair in the front room of my house. One winter three years ago, 17 of my haircuts died. You know, at three dollars a haircut, that's fifty-one dollars a month. Now I have only about 16 customers left. They're all pretty well dead.

I am 75 years old. I am rich. I don't need any more money. I've got enough money to do a lot of things. I drive a good car. I'm going to stay here. What the hell else can I do? I am 75 years old. Where in the hell am I going to go?

ALEX RONYK

I came here from Brandon, Manitoba in 1935. I came during the hard times and I couldn't find a job anywhere. I worked a couple of months, then went back to the farm in Manitoba.

But I came back in '43 and have been here ever since. My first job in the winter of '43 was driving truck. In '49 and '50, I fired a locomotive. I shovelled the coal and you really had to shovel when it was going up a hill. I skinned cat and worked around the tipple as well.

But mainly, for 20 years, I was on the dragline. From there, I got into management. I suppose dragline operating was one of the most prestigious jobs in the field.

The first dragline I ever ran was powered by a diesel motor. On those early draglines, you didn't have any help on your controls. They were all mechanical and very hard to pull. The levers were long to give the operators a bit more leverage. Those equipped with air valves for controls were easy to operate. When the draglines became electric, it took no effort to operate them, just no effort.

I never did get injured but – I suppose it was lack of exercise – my circulation got bad. It got so I could hardly walk. I had an operation and went back to work for one year. Then I went back on disability pension for the last few years.

There have always been two or three unions in this coalfield. I was always a member of the union until I became president in 1969 to 1972. It was a thankless job. – Stewart Kidd

You know, I feel good except for my arthritis – it's bad sometimes. I can still get around, so I want to stay in East Coulee. I was born in Estevan in 1919. My dad was a coal miner in Bienfait during the winter and a farmer in summer. I met my husband, Toby, at a dance in Bienfait. He was a coal miner. We got married in Estevan in 1942 and later moved to the Drumheller Valley in 1948.

At the time, the underground coal mines were closing in Saskatchewan. Toby wanted to continue as a coal miner so he got a job in East Coulee. His brother was already working in the mines there. We weren't going to stay long – just enough to save our money and move out to the coast.

In East Coulee, I worked in the lumber yard. We had a big general store on the corner and I also worked there. I had various jobs until I wound up with the Atlas Mine in 1972. They hired me to weigh trucks. I ended up doing the shipping, payroll and books – I also sold coal. I worked until 1987 when the owner, Omer Patrick, donated the tipple, associated buildings and land to the Dinosaur Valley Heritage Society.

I think if the mine had been operating on a small scale, I would still be hobbling over there. I really did like the work and there were many nice people. – Julie Auld

I got a nice garden and lawn. I pretty well had a garden all my life. See, those are all fruit trees out there. I like to tinker around with stuff like that. Well, it's something to do, especially since I retired, there isn't a helluva lot else to do.

I was born in Taylorton, Saskatchewan, in 1911. My dad ran a store in Taylorton and did a little rum-running across the border. I started coal mining at the M&S mine in Bienfait, Saskatchewan. I worked about 40-odd years – all of it underground. I was a contract miner at one time and from there I went on a loading machine. After that, I wound up an underground mechanic. I was jack-of-all trades, fire boss, mechanic and joy-loader operator. And I spent five years in the army.

When we moved to East Coulee from Bienfait, we couldn't even get a place. It was full of shacks. I think there was somewhere around 400 miners there. It was a damn good place – by God! I think there were six stores and about four beer parlours.

I think I have overlived my distance, right now. I think when you get over 70, you're on borrowed time. – John "Toby" Auld

ROSE AND STEWART KIDD, RETIRED DRAGLINE OPERATOR – BIENFAIT;

JULIE AULD, RETIRED OFFICE SUPERVISOR & **JOHN** "TOBY" AULD,

RETIRED COAL MINER – EAST COULEE, ALBERTA

EARLING ROTHE

BOB TOOMBS, ENGINEER & **GARY** GEDAK, BRAKEMAN

COSTELLO MINE

LAWRENCE CHARLEBOIS

ELMER HOLSTEINE

SHIFT FOREMAN – COSTELLO MINE

CHAR BRIQETTE PLANT

LES KINGDON

I was born in 1893, in England and came to Canada in 1914. I worked underground for 20 years in various mines in Bienfait, including the Manitoba and Saskatchewan Coal Company. Later, I operated my own underground mine near Bienfait and operated it for 15 years. Both my sons have been involved in coal mining in Saskatchewan and Alberta.

When I came to this country, I had no training or schooling. My dad was a miner so he asked the boss to give me a job. I started in 1922 – I had no chance to learn anything else. When the war broke out, miners were soldiers and they couldn't move to another job. That's how I spent 50 years of my life.

You know, it's interesting that it took 50 years to realize that the big strike that took place here was important. Last year was the 50th anniversary of the strike. I spent a lot of time helping people learn about the Estevan Strike. They wrote a play about it and the TV guys came here.

In 1931, that was the first time the miners went on strike. It was the first union to come into Saskatchewan. The mine operators were against the union – they would fire you if they found out you were a member. The union finally got organized and made a proposal for better working conditions. The owners were exploiting the miners as much as they could with low wages and short-shrifting on the tonnage of coal mined. The owners would not agree to improve conditions. They called the union members "reds" and trouble-makers.

When they went on strike, the government sent a bunch of Mounties down here. The Mounties were stationed in Estevan and they used to ride around the camp with a truck that had a machine gun on it. It was just to intimidate the miners even though the miners hadn't done nothing.

Finally, the union went to Estevan to protest that the city fathers were sending the Mounties down to intimidate the miners. When the miners got close to the town hall, they were stopped. They scattered but the Mounties started shooting. Three miners got killed and several were wounded.

I happened to be in the first truck to Estevan and here were the Mounties who wouldn't let us go any further. The union leader was an Englishman and most of the miners were foreigners – they couldn't speak well. The chief of police came over to our truck and started arguing about various things. He got hold of our leader and dragged him off the truck.

The truck was low and he got a hold of his head and was trying to pull him off. Of course, the miners got a hold of him and wouldn't let him go. They were arguing back and forth and pulling. Then one of the miners gave a helluva smack to the chief of police, knocked his cap off, and that's when the whole thing started. The Mounties and the police were backing up towards the town office and the miners were just walking towards them but when they got their backs to the wall, they started shooting.

We took one guy who'd been shot to the hospital but the nurses and doctors wouldn't treat him – called him a "red bastard." We had to drive 50 miles to Weyburn to get him treated but he died. That was a black Tuesday. After the incident, a Royal Commission was set up to investigate the shooting. They arrested the union leaders and sent them to jail. It took until 1948 before the UMWA came back to the Estevan area.

After the strike, I couldn't get a job nowhere. So I had to work in the small mines around the area. I went to Mercoal in Alberta and worked in the mine there but then I came back to Bienfait and got a job in the small mines until 1940, when the war broke out. There was a shortage of miners and I was experienced with machines. The manager who'd called me "red" and had fired me came to see if I would go to work.

Finally, I went back to Western Dominion. I was there until they finished that mine. From there, I went to the Utility Mine and worked there until I retired.

When I think of it, a lot of coal miners went through misery. Cheap wages, no air, all that stink, all that crap and sweat. Now I think, how the hell did we live through it? It's just like a bug when he gets on the horseradish, he thinks there's nothing sweeter than the radish. Miners are the same way.

I was 15 years old when I started and I could make a dollar. That's all that matters. When I started, I couldn't read or write. Now I can read and write in both Ukrainian and English.

RETIRED COAL MINER

PETER GEMBY

was 21 when I started working for the Winnipeg Fuel and Supply Company at their strip mine here in Roche Percee. I started on the tipple as part of the clean-up crew. Later, I moved to the Old Mac Mine and then finally to Costello Mine. Mainly, I worked on the tipple, loading rail cars.

During the first few years, I worked at the mine in winter and ranched in summer. I preferred the cleaner life of ranching. The tipple was awful dusty, noisy and it smelled of oil sprays.

We were all in the mines for the dollar but it had its good points. It was always good for a laugh a day and you met some real different characters.

Now that I'm retired, I'm just taking it easy and enjoying living here in the valley. — *Robert Long*

was born in England, but we came to Canada in 1920 when I was six months old. My dad was a farmer.

When I came back from overseas in 1945, I had nothing. I figured I'd get a job for the winter. Like one fellow said, "Don't stay here too long!" Because once you stay here a year, you never leave. And he was right because I never did leave.

I worked in Roche Percee for the Winnipeg Fuel and Supply Company until 1961. I started out in the soup kitchen, cleaning out the sump hole. I worked there for three days. I guess they put me there to see if I would stick it out. Then I started loading boxcars with the lump loader.

Then I went up on the tipple as I worked my way up the ladder. In those days, you started out as a labourer. I put 23 or 24 years in the same job and never did move around much. I could have been on the dragline but I never did want to.

You're the first one to ever come around and say he wanted my picture. It's nice to have someone interested in the old fellows — guys working in coal mining. They say, "Oh, he's just a damn old coal miner" and they'll put him in the ground and forget about him.

I'm a farmer now. The other day, I was riding a combine and a skunk came out from underneath the swath right in front of the pickup. It was a variable-speed combine and I got excited and I pulled the variable speed out and the goddamn combine picked up speed and, Jesus, it caught the skunk and that son-of-a-bitch went through the combine. I am going to tell you about stink. Holy old lightning, you couldn't stay on that son-of-a-bitch. Your eyes watered and you could even taste it. It gets in your clothes and everything. — *Stan Long*

ROBERT LONG & **STAN** LONG

SUE AND BURT STOCK

I was born in 1914 on a farm just 32 miles east of Bienfait. I was about seven years old when we moved to old Taylorton. My father worked in the Sugar Loaf Mine as the pit boss. In 1927, there was something like 35 mines around here.

I started out in 1929 at the Old Taylorton Mine as a trip rider and I got seventeen-and-a-half cents an hour. At that time, you could buy a pair of good overalls for 75 cents. I worked at that mine until 1936, when it burnt down. From 1936 to 1940, I worked in various small mines such as the Old Crescent and East Collieries. Then I went to the M&S Mine and stayed there until I retired in 1969.

Actually, the times have not changed that much. The only change is the working conditions. You have a lot better working conditions now than you had then. When I came back from the army, they asked me if I wanted to work on top or down below. I said "down below" because I preferred it — it's always the same temperature down there. Also, you were closer to your fellow workers when you were down below than on top.

I remember a fellow who used do rum-running and bootlegging pigs. He was bootlegging pigs because they were next to nothing in price across the line and there was a pretty good price over here so he was bringing them up by the truckload.

One time, he just got across the border when the Mounties took off after him. He had this dump truck so he just threw it into gear and dumped all the pigs out. When the Mounties caught up with him, he had no pigs or nothing in his truck. The next day he went around gathering up those bloody pigs. Oh, there was lots of fun in those days. — *Burt Stock*

SARAH AND SAM LESTER

I was born in Roche Percee in 1918. I worked for Manalta and Roche Percee 31 years. Overall, I worked in those underground mines 15 years.

Well, I did just about everything at the mine. I liked underground better than surface – oh hell, yes! But I sure miss those shovels. The manager and I got along well. One time, I bought an old truck from him. First, he said to me, "You're not getting that truck." I said, "I'll give you eighty dollars but you have to take it in and put a battery and new spark plugs in, set the points." He said, "You're a robber." So him and I got along good.

We lived on the farm and had about 12 cattle. We finally had to sell them because she had to walk down there and chop the ice. She used to fork the hay. Finally, we found an old bailer. She used to sit behind the bailer and tie the bails. Lightning and rain, we kept right on going.

I retired in 1981 after a heart attack. – *Sam Lester*

GLADYS AND JOE WRIGLEY

I went to school here in Roche Percee. This is where I learned howling and fighting and not reading and writing. There were six mines here and they used to raft the coal on Souris River to get it to Winnipeg. Roche Percee was once much bigger than it is now. They had a hotel, two stores, a pool hall. Now there's no store – not even a post office.

I was born in England in 1912, the same year we moved to Canada. My father was a coal miner in England. When he arrived in Canada, he worked underground in the Old Taylorton Mine near Roche Percee. He later worked for the Manitoba and Saskatchewan Coal Company in Bienfait and helped build the M&S Tipple 75 years ago.

When I was 14 years old, my first job was hauling coal on surface with a team of horses. I worked underground in mines at Bienfait and in Red Cliff and Drumheller, Alberta. In 1932, I went to the M&S Mine in Bienfait and worked there for 42 years, firing the lignite boilers. In later years, I worked in the briquetting plant and was a watchman at the mine.

I enjoyed working for M&S. Working underground was a job whether you liked it or not. I retired five years ago. I always tell people, that's when they fired me.

I am in the process of moving to a retirement home in Estevan. It's going to be sad to leave my large garden. I just put it in. You know, the valley of Roche Percee is beautiful. – *Joe Wrigley*

COAL MINERS

OF NEW BRUNSWICK

Chipman

Minto

NB Coal

Grand Lake

Newcastle Bridge
New Zion
Newcastle Creek

● PHOTOGRAPH LOCATION   ■ COAL MINE LOCATION

DAVIE BETTS, **FIRMAN** BETTS, **DEBBIE** THIBIDEAU & **GLENN** BETTS

I get more money on the pension than I ever earned in the mines. I'm getting so damn old that I forget everything that happened to me. But I still go fishing every day.

I was born in 1900 and came to Minto in 1914. I went in the mine the same year and worked 39 years underground. I wore out a lot of carbide lamps in my time.

I had 13 kids. There were some hard old times from '22 to '39. She was damned hard. We had some good storekeepers here, by gosh. They would give you credit if you got stuck. If you were sick, they would carry you until you got work. I worked a total of 54 years and retired in 1968. The last five years, I worked down here in the shop.

I liked the mines. You'd be out of the cold, out of the heat, but you'd damn near freeze going home after you'd come out. They had no changing shack – just a place to go into the boiler room to eat. They got washrooms in after I quit. – *Firman Betts*

STANLEY CRAWFORD

I was born in 1904 in New Brunswick. When we built our house here 43 years ago, this was called Shaft Road. There were shafts all around here – you could see them sticking up through the trees. There was a shaft 200 feet from here and another down at Bram Moore's. The coal was 42 feet deep right here. A rail line was put into pretty well every shaft. At that time, it was the only transportation, other than horses.

Back 40 years ago, there were hard times in Minto, I can tell you. Oh, I guess there were hard times everywhere. I started at a small mine in Salmon River where you had to work and hope for your money. Sometimes you worked and didn't get a paycheque and, of course, the paycheque you got wasn't too good.

You'd be working for $2.70 a day, one or two days a week. You tell me what you'd do! Of course, you'd do what I done – you'd just go on credit.

We had eight children. Like I said, when the wife set the table for a meal, she set down 10 plates. I wonder how we ever got them a bite to eat. But we had lots of fun as young people and they all grew up quite rugged lads. I was active in the United Mine Workers. Like I said, when we first joined the union, we had lots of strikes. Sometimes you'd gain by a strike and sometimes you wouldn't. We had one strike here for nine solid months.

You don't forget those times, because you had no money to fall back on and the storekeepers would only carry you so far.

BRAMWELL MOORE & GRANDDAUGHTER, TAMMY MOORE

A fellow has to do something. He has got to work. He can't be idle, laying around in a rocking chair, I'll tell you.

I was born in Yorkshire, England, in 1904 and came to Canada in 1913. My father was a coal miner in Yorkshire and when we came to Canada, he worked underground in Minto.

My first job was wheeling when I was 14 years old. I retired in 1969 after 39 years, all of them underground in the Minto area. I had been wheeling – that is, loading the boxes and pushing them out for the miners – and, anyway, I decided I was going to go into mining for myself and dig coal. I got an auger and picks and I asked the boss for a site. I got a site and started digging coal.

The first day, I dug six boxes and those boxes would take about two to make a ton. Boy! I was making money. The next day, I got four. I went for a couple of days and got two.

Then the place caved in on me. So I took to wheeling and from there I went back to digging again and finally got onto it.

I got six boys. My boys used to work with me when school was over. They never thought much of that. None of my boys took to mining – they all got out of it. It was just as well as it was phasing out for the draglines, anyway.

My dog is the one and only pal I've got. His name is Chapnab.

BILLY SPENCER

I was down in the mine when I was nine years old. That's when I first went down. I've been in the mines about 37 years altogether. My dad worked in the mines a lot but he was kind of a foreman. My brothers worked underground and I took my boys underground.

According to Frank Coakley, "Billy was a hard man, boy. He ain't very big but, I'm telling you, it took some kind of man to overcome him. He appeared to enjoy fighting. He went to Cape Breton one time. He was telling me that he was going down for a week with his buddy to work in the mines. His buddy said, "Now, Spencer, keep your damn mouth shut down there or these Cape Bretoners will kill you." Billy said, "I know they'll kill me," and, the first night they were there they got into a fight and there were none of them Cape Bretoners could do a thing with him. "Quite a lad, Billy!"

RETIRED COAL MINER

FRANK COAKLEY

I used to play the mouth organ and the Jew's harp. But this town here – I don't think all across the goddamn Dominion of Canada that there is any more talent than in this town. You'd just be surprised. I imagine there are at least eight bands in this town – damn good ones. I still go to the Hootenanny every Sunday night to dance and watch the girls trying to attract attention.

I was born in 1904 in Saint John, New Brunswick.

I don't know much more about coal mining than the rest of them. What I do know is that I tried to scratch a living at it while I was doing it. I didn't pack it in when the Depression was on because there was always a day's work here in Minto. They'd come clear from British Columbia to scratch a day or two of work.

You can go all over the world and find some soul who's been to Minto and done some sort of work. Well, I never cared too much for digging coal. If I had my life to live over again

– of course, you know it's too late – I'd never go down a coal mine again. Oh, the young fellows would say, "I wouldn't do work like that." Well, if there's no alternative and your stomach is empty, you do anything.

I didn't mind working on the machines when they started cutting the coal. That was okay and the air was pretty good. But it was tough when you were using the hand pick and shooting all the time. The smoke from the powder would give you a headache. You had to go to bed lots of times without your supper.

I didn't put much more than 20 years underground. I came out in 1948. Then I worked on the strips until 1968.

The kids today are having a better life than when I grew up. See, when I grew up, there was no money. If you wanted to go anywhere, you had to walk. Today, all the kids got cars. They all got money.

DICK QUIGLEY

y name is Irish. I was born here and my dad was born in Nova Scotia.

There were eight of us in our family — six boys and two girls. Pretty near all of us worked in the mines. My brothers were foremen. One brother got killed in an explosion and he was a foreman, too. The underground explosives caught fire and he tried to get the men out but he got killed. You see, it wasn't all good.

I worked about 30 to 35 years in coal mines. I just quit because I figured it was going to end, anyways. I was four years with Department of Mines as underground safety officer.

Mining coal is quite a thing because you get to do everything yourself, from drill work to explosives and all your own timbering. If you had a handpick mine, you got to do all your own work. It's up to you to see that you don't get buried. You got to watch what you're doing.

It's all strip mining now. There are no shafts. But if they had their shaft and a good seam of coal, they would get the men. A lot of young men would be interested in coal. Coal mining is quite a thing.

Oh yeah, that's where I do all my walking, back there in the bush. I get a little wood, you know. I don't want to die yet so I get a little exercise by keeping myself moving.

RUSSELL ACKERMAN & ARTHUR ACKERMAN

ARTHUR: We were both born in New Zion, across the road from here. Father worked around the mines on surface, shovelling coal into boxcars. We got started by wheeling with our neighbours – we both wheeled for quite a while.

Yeah, we both started when we were 13.

RUSSELL: I stayed underground right to the last year, then I worked on the picking table for a few years.

ARTHUR: Twenty years ago, I stopped working underground. Then I worked on the draglines and did carpentry work. I repaired the washplant and built powder houses. Five years ago, I quit.

RUSSELL: At the time, I enjoyed coal mining. Yeah, it was better than outside work. You could do your day's work, then you were done. No one jumping on your back. Sit down five minutes on the top and they'd jump right on your back. You could go underground and if you worked two or three hours, then you could sit down for an hour and talk to people.

ARTHUR: If I was 18 or 20, I'd go back to the mines. We never went high enough in school. Now that we're retired, we cut a lot of wood and have a big garden.

FREIDA AND CLARE ROBERTS

I know some of the old-time songs. I often go over to Louie Madore's and sing while Louie plays the fiddle. We went to the Hootenanny last night and it took me nearly all night to get my feet going. Today, I'm having a hard time getting my brain working and I might not be able to sing any songs for you.

I worked 20 years underground and nearly another 20 years on surface around the draglines. I was working underground for one guy and the son-of-a-bitch threatened to fire me. You see, if you had a roof fall, you had to clean it up and you didn't get paid. You were paid between 25 and 27 cents per box of coal mined but if you were unlucky and had a roof fall, you wouldn't make any money that day. So, if a guy had bad luck, you'd give him a couple of your boxes so he'd get paid. Anyway, the manager found out we were doing this and threatened to fire us. Christ, you had to work all winter to get a day's pay and then, half the time you worked, you didn't get paid for it. Nobody cared whether you died or not.

The manager was an old hillbilly. He stopped me one day in town after I'd had a couple of drinks. He said, "You fell off the wagon, didn't you?" I said, "You look drunker than I am."

We had a lot of fun when we was working, but the manager figured everyone was a thief. Gasoline used to disappear very fast. Gracious God Almighty, that's what's wrong with this country. If you do get anything, the owner will come and take it back.

Since we quit work, we don't go to bed early and we don't get up early. We always watch the late news. We sleep in a bit in the morning because we have nothing else to do. – Clare Roberts

JIMMY FORRESTER

I have a little greenhouse back here where I started some of my produce – tomato plants, cucumber plants, etc. It's just a hobby. I don't go into it heavy enough to make it worthwhile to start peddling produce. One fellow said, It's no wonder I have a good garden because I live on it, and I said, Yes and a little later, I will be living in it.

In 1907, I was born in Glasgow, Scotland. When I first came to Canada, I worked as a farm hand along the Saint John River valley. It was the very early '30s when I decided that the fellows working down in the pit were making money. They'd be having their evenings off and I would be out haying and bringing in the cows to milk and one thing or another. That's when I decided to make the switch.

My first job was wheeling for an old Hungarian. My first day in the mine, I was buried and my partner was killed. It was an hour before they had me dug out. I was off work for months because I broke a number of bones. I recovered enough to go down the mines to try wheeling again, but it was too much. So I got a job shovelling coal in boxcars.

After the war, they started to dig with an old steam dragline. I started as a pumpman, then I worked around at various jobs. When I finished, I was a loader operator. I was really enjoying surface mining when we were laid off in 1969. All my work was with the Avon Coal Company.

ROSE AND MANUEL CANTINI

MANUEL: I was 24 years old when I came from Italy. I was all alone when I came here. I got married here. I am 89 years old.

I worked in the mining industry for over 40 years. There were quite a few Italian coal miners in Minto. Not too many Italians are left now but at one time we were a large, close group. She is French and me Italian.

ROSE: I can talk it but I am not Italian. My father was a coal miner in Minto and in the old country, France. I was born in Calais, a coal-mining area.

MANUEL: I was very glad when I retired.

This was a nice garden a few years ago. It's hard to work now because I get tired. Working passes the time a little faster. A person can't stay in the house all day. He's got to get outside, just to do anything at all to pass the time. If a person do nothing, that's worse. Even in the summertime, from here I walk to Minto. Not very often but once in a while. I like to take a little walk.

I went some other place but I never found a better place than here. I rather be here than in some city. I rather be outside and work on the land.

JIMMY RICHARDSON

don't know how I done it when I look back now.

I started when I was 16 and worked 28 years in coal mining. I started wheeling for my dad when I was 16. When I was 18, I got my miner's papers and a place of my own.

If I hadn't got my back broke when a cage came down on top of me, I would have kept on mining. I really enjoyed the mines. I never worked on surface, always underground.

We had bad places in the mine. What I mean by a bad place is a wet place. You had water dripping out of the roof and that's bad. Water on the floor wasn't too bad. Water overhead got you wet all over. Down in the mines, it's cool. If you stop to have lunch, you get cold, but when you had a nice dry place, it was good.

Well, the mines made us a living anyway and I brought my family up. Everybody had four, five or six children. And our family, they all went to high school.

GAIETY THEATRE

ED RYAN

I was born in Minto in 1915. My dad was a coal miner.

I was about 14 years old when I started in the mines. Starting in coal mining was very simple: if you wanted to eat, you had to work, that's what. I started oiling boxes at a shaft for the Minto Coal Company – I got a dollar a day. I didn't work there very long, then I went wheeling. After that, a fella just graduates into digging. I've done just about everything in a coal mine, I guess. I even helped dig a few fellows out who got buried.

It surprised me that I worked in so many shafts. The shafts weren't deep and they had to move them every so often or sink a new shaft. Instead of making haulages to haul the coal, they would just sink another shaft about 3,000 or 4,000 feet away and start all over again.

I retired three years ago. When I retired, I was hauling coal – trucking. You make more money trucking coal, but I kinda missed underground. Trucking was easier and you were in the fresh air all the time. There's something about the mines: once you've worked in them, you miss them. One thing: the guys who work in coal mines are a great gang.

When they had all the shafts here and the different companies, you could always get a job somewhere. Take a good miner, he never had to worry about his job.

I worked 58 years in the coal industry in Minto.

MADELINE FILA & **GARNET** RICHARDSON

I was born in 1911 in New Brunswick. My father came from Nova Scotia. He spent his life in the mines.

I was eight years old when I started and I worked in the coal industry for 55 years. When I was young, I used to work on the dump after school for a piece of cake.

My dad took me down the mines. All my schooling was in coal mines. The last few years in the mine was a pleasure compared to what it used to be years ago.

It was a pleasure because if your place had bad air, they put a fan into it and if the place had water, they had to drain it or put a pump in there. In the old days, you had to lay right down in the water 'cause if you didn't do it, somebody else would take your place. Fishing is my main thing now. I go way up the Miramichi. – *Garnet Richardson*

FRED LEGERE

In 1908, I was born in Joggins, Nova Scotia. My dad worked quite a few years in the mines and he worked up here five years and in Springhill and Joggins. I worked with him at River Hebert Mine in Joggins for six months.

Then everybody said Minto was booming. So Dad said, "Let's go to Minto" and a whole bunch of us came to Minto in 1923 and we stayed here. I've got 46 years in the mine, all of it underground.

I dug most of the time and ran coal machines as well. I retired when I was 60, then I worked three years after that.

Every miner who has 35 to 40 years underground has coal dust in his lungs. Anybody who's worked around a coal machine, got lots of dust. I had two brothers who also worked in the mine. One brother got his eyes blown out in the mine.

A shot blew both eyes out. He was 29 years old.

DOMINIC DiCARLO

I was born in 1894 near Rome, in Italy. When I was 14, I looked after the sheep for my father. Instead of beating the lambs, I played with them and they grew nice and big and I got a better price for them.

In 1912, I came to Minto from the States. I met a fellow who told me about Minto but he said that he didn't like it because the coal was only 18 inches high. When I came to Minto, I worked for Minto Coal Company and a few small mines.

I bought a store right away. I had my first store at New Zion, near the vault, up where they put the dead people. I sold that and bought a store here with a barber shop and restaurant.

We had a fire and I rebuilt it – bigger. I worked quite a few years in the mine. Even after I had the store, I worked in the pit because I was pretty good and make money. I make more money in the pit than in the store. Later, I owned a mine and ran it for five or six years. My boys worked in it a bit.

Oh yes, everybody know me. Come lots of people – not like your size – but the kids come and play with me.

We had lots of fun. I work in the store everyday. Oh yes, I never lost a day.

ELI GOGUEN

I was born in Boston in 1892 and worked in the coal mines in the Minto area for about 50 years. When I first worked in the mines, we had no electricity so we had to use pails to bail the water out. In the old days, we used to wear oilskin coats, just like the fishermen. At first, we didn't have fans to clear out the dead air, but there wasn't a lot of coal dust like in Nova Scotia.

I worked in the mine most of the time and I did various jobs in the mine. At last, I got to be boss of somebody else, looking after the cutting crew. I didn't want to boss anybody around, so I went back behind the machine, timbering.

I raised some chickens and had some hens, too. I kept horses and I kept cows. During the Depression, Minto was the only place in New Brunswick with a dollar. The rest of it — there was no money. Here in Minto, I've seen lots of times in the summer when we had a day or two of work a week. And if you had some cows and a little garden, it helped some. In the winter, I used to get a lot of moose and deer meat.

I used to live in one of the company houses around the mines. There's still company houses in Minto but they're now remodelled and owned by the people who live in them.

QUEENI AND BILL BUSCH

I was born in Germany in 1903 and came to Minto in 1913. In 1914, we – my mother, my sister and I – used to go down with my father at night and help him shovel the mud back while he was digging the coal and driving the lath ahead. I had a little shovel. My sister would cry and yell because she was three years younger than I was – but mother, she'd help shovel the mud back. Oh boys, oh boys! Years ago, a lot of women used to go down the mines.

All through New Zion there were many mines. I worked in every goddamn hole up this way. When I didn't work for my father, I worked by myself or wheeled for somebody else.

If they could hit about three or four feet of coal here, I wouldn't kick because I would go back digging tomorrow. You're blessed right! I'd love to know if there's more coal underneath Minto. This was a prosperous little town at one time. There were over 1,000 people just mining and now, my goodness, it's nothing like it was.

Now, a dragline is a great thing for the moneyman but it's no good as far as labour is concerned because you can run a dragline with 25 or 30 men and you got 300 or 400 people on the streets. And the price of coal is still the same.

As the saying goes: "Give me a small cup one day and a large cup the next and we'll all be working steady."

I remember the day Fred Richard got buried. There were 15 men there and not one of them would go and help him out. Well, I went in and crawled over the mud and scraped the mud back. I told him to keep moving the mud away so I could get near him. He was packed right into a corner behind the prop with his head between his legs. I worked an hour and a half to get him out – by God, I was tired out. I worked like a horse to save Fred. – *Bill Busch*

RUBIN REED

I was born right here in Minto in 1914. I was 13 years old when I started in the mines. Then I went to Montreal for quite a while and then I was in the Merchant Navy. All the rest of my time was in mines and draglines. After the underground closed, I went on draglines.

The dragline was all right. If you were running, you got lazy. The trouble with the draglines was that you had to work eight hours. The trouble with most of us here, we drank a lot and had good times, too. But when some of us got to a certain age, they wouldn't quit. They tried to keep going like when they were 20 or 30. Those guys are all gone now.

Well, like I say, if I was young again, I would go back because I don't think it's that bad a life. Oh yeah, there were hard times, depending on what level you had. But it had a lot to do

with you, too. If you were working in a place, you wanted to timber it right and keep it clean and keep your levels right. It wasn't that cold and wasn't that hot – it was about 35 degrees all the time. Oh, I liked the pit. Oh Christ, yeah! If you wanted to go home at noon, you'd go on home. And there was nobody at your back watching you work to tell you what to do.

I had a big family – about nine – and I kicked them all out. One boy worked in the mines but he only lasted a couple of weeks. I figure there's no future in it for the young people. But it was okay for us. A kid has a chance today. You can always get work in Minto – this is one place where you can get a job any day. In my life, mister, if you didn't work, you didn't eat!

WENDALL WELTON

L ove of the land, I've been drinking since 1922, that's when I was born. How could you travel in this world and find a nicer guy than me? I was born in Minto in 1922. My family is one of the oldest in the Minto area. My father mined coal for my uncle, Harvey Welton, who owned a couple of mines in the area.

I started wheeling when I was 13 years old. After that, I joined the Merchant Marine and travelled throughout the world. I was 17 years old when I first visited Calcutta.

When I came back to Minto, I worked underground at a number of mines including Newcastle Coal at Black Diamond. Eventually, I went on surface with the draglines and worked as a lineman and cable-splicer. In 1987, I retired from N.B. Coal.

I beat the best guy in Fredericton, but I'm not a fighter. I never got all these wrinkles from laughing. I was a bad bastard.

LOUIE MADORE

was born in 1915, in Nova Scotia and we came here in 1920. The summer I was 12, I stayed out of school and went into the mines. When I was 13, I went in the mine and stayed there.

You take years ago when we came here — it was all underground mining. There were coal mines down in this area (Avon Settlement). Can you see that slag pile? That's from the old mines. All around this area were shafts and them slag piles of coal rock from the mines.

One time, many years ago when old South Minto was going, there was everybody here from under the sun, every nationality. It was a rough place. There were no police here for a long time. But it simmered down after a while. We went out to Canmore in 1969 when they started phasing out the mining here. I was there from '69 to '75. The mining methods were different but there was a great comparison in the people. There were all the nationalities in Canmore, the same there as here.

We worked all our lives with 16 to 18 inches of coal when out there, it's 18 feet. They tell me when they closed the mines in Drumheller, there was a three-foot seam and the miners would not work in it. They should have sent work to Minto because at that time, the boys were working in 16 inches of coal.

Around the area, there are some great old-time fiddlers, you know, not just diddlers but musicians of all kinds. Really good! Last year, my nephew and I played in an old-time fiddling contest. I don't play enough to stay in practice.

Oh yeah, we had coal mining songs. The old timers, they were some good. They used to make up songs about you and sing them to everybody. One song was made up right here by people in this district — it goes like this:

> *This is the place where the coal miners gather*
> *There's pit lamps, there's gumboots*
> *There's clothes scattered around*
> *You work like the devil down in the bi-level*
> *While putting out coal in the coal-mining town.*

HAROLD BROWN JR., **ALVIN** BROWN & **HAROLD** BROWN

I worked in the mine pretty near all my life. I'm young yet, of course – I'm only 82.

I was in the first war. When I came back from the war, I went back in the mines and I stayed in the mines until I retired. I started in 1921 and quit when I was 60.

I used to drive the main levels. It caused a lot of sweat because you were always in solid ground. It was so wet in there that when I came home at night, I could still hear the water running.

I liked digging coal. When you went in there and did your work, then you could go home, regardless of what time it was. If you were a fairly good miner, you got home early. If you were a poor miner, it took you all day.

I was in the Halifax explosion. I was in the cookhouse and had a broom, sweeping the floor, when the explosion happened. It got so dark you couldn't see your own hand. Something hit me in the knee and I was in the hospital for a month.

This railway has been here since 1900. They used to have the Cannonball here that would go to Saint John and back the same day. The midnight Cannonball used to come in at midnight. There was quite a big station at Minto and another at Newcastle. They had a train called the Sulphurball Express.

I don't know nothing but diggin' coal and cuttin' wood! I split that pile of wood last winter for exercise. – *Harold Brown*

WELLESLEY HOYT

started from the bare, bare bottom and worked my way up to the top in both strip and underground mining.

There's not a job in the strip-mining field that I haven't done with my own hands. Our name is of Dutch origin. My father was born in New Brunswick. He worked in hardrock mines and became a self-educated mining engineer in Texas.

He came back to New Brunswick and started an Antimony Mine near Fredericton. In 1927, he began coal mining for the Miramichi Lumber Company in Minto. He was here until he died, eleven years ago.

SURFACE COAL MINE

HARRY THOMPSON

It's been a long time – over 45 years – since I started working in the coal industry.

I was born in 1919 at Hardwood Ridge, near Chipman. I started out as a boy, 16 years old, working wheeling for a fellow down at Newcastle Bridge. My training didn't amount to much – I went through grade 8 but I completed high school arithmetic. After I started working in the office, I took a course in accounting.

Then I went to Avon Coal Company. Did everything underground at that mine until 1939. After about four years underground, I got a job working in the office for a small coal operator down at Coal Creek. Two years later, I went to work for the Minto Coal Company as a general clerk and shipper. I stayed there until 1946 when they sold out to the Miramichi Lumber Company, one of the large producers in Minto. From there, I went to DW & RA Mills as a general clerk. A few years later, when the office manager died, I took over his job and held that position until 1969 when they stopped private mining in Minto.

During the time I was with Mills, it was pretty interesting. In the beginning, we were a small company and were using an old steam-operated dragline. At the end, we were the largest company and had the largest operating dragline. But it was difficult to make any money. You could only get about seven dollars a ton for the coal and the dragline had cost us more than two million dollars. We had our ups and downs.

At the time I worked for Mills, I was in partnership with Alton Hoyt in a small surface mine over at Coal Creek. We produced 7,000 tons a months and sold it to the Power Commission. In 1969, when the provincially owned N.B. Coal took over the coal operations in Minto, I became the comptroller and I still hold that position.

AUBREY ELLIOTT, MAINTENANCE FOREMAN & **ART** THOMPSON, WELDER

NB COAL

BILLIE FORRESTER

WELDER – NB COAL

DWAINE BARTON

WELDER – NB COAL

ERIC BARNETT, **HAZEN** BROWN, **WALTER** FULTON,
CAMERON KNOX, **ELI** RICHARDS & **ELVIN** WOOD

WELDERS — NB COAL

VAUGHN LEGASSIE & **HAROLD** CAMPBELL

ALTON HOYT

was born in 1928 in Minto. My first job was on the spoil bank, shovelling coal. After that, I decided to go railroading, so I worked on the CPR as a fireman for about four years. I didn't like the shift work so I came back.

At that time – it was 1961 – the shaft at the Newcastle Coal Mine burned and they said it wasn't going to open again. This meant there would be no work for any of the miners. One of my older brothers came to me – he and my dad were interested in underground mining. I was asked if I'd be interested in buying the Newcastle Coal Mine because the owner was wanting to sell. I discussed it with the old man. He said, "Boy, go ahead." So, with the money my wife and I had saved, I bought the company out. We rebuilt the entire structure and were producing coal in three or four weeks. That's how it came about. It was a small underground mine on the Post Road.

Once the mine got under way, we worked seven days a week for six or eight months, fixing things. Any extra work we did ourselves because we just couldn't afford to hire anyone. So we let it go to weekends and just did it ourselves until we got to a position where we could afford to have shift workers come back on the weekends.

Then we lost our contract with the power plant. That's how I began to get involved in politics. I didn't realize what politics were about until then. We were in competition with other mines in the area and the crunch came when our competitors started buying draglines and other large surface-mining equipment.

In 1963, Harry Thompson and I decided to go into stripping at Coal Creek. We formed the Midland Mining Company. In 1965, we were getting $9.20 a ton and had two mines. Our last mine – Shaft No. 7 – was closed in 1969. Now I'm mining fine coal with a dredge on the old tailings pond.

CLARENCE CRAWFORD & **ROY** SONYIER

ROY MILLS

I always felt that if I didn't try mining, I would always regret it. The first strip mining took place in the Minto area about 1910. They used scrapers to do it along the Fredericton Road. The owners had heard about these steam-powered draglines that my family had working in Quebec.

So then we came here from Montreal in 1944. We actually started in Minto for the Bank of Nova Scotia which owned the Minto Coal Company. My brother Dave and I decided to stay here in Minto to develop our own coal mine. We worked as contractors and, when the Minto Coal Company was sold, we continued to work for the new owners, the Miramichi Lumber Company.

My wife's father sold the coal for us. He had a lot of connections because he was the former manager of the Minto Coal Company. We kept getting bigger and bigger. We used the old draglines from Quebec for a long time and eventually bought some new ones. We got up to 250,000 tons a year — oh yeah, we produced a lot of coal.

After we were bought out by the province in 1969, I was retained a few years as manager of the N.B. Coal. Then I retired. We had a big family — nine children. So, after I retired, I spent the time visiting the family. But we got fed up with that so I got back into the business.

We kept the land with the tailings on it from the old mining washplant operations. Our thought was that if the price of coal were to increase, we could develop the tailings that contained up to 40 per cent coal. There is a block of a half-million tons of tailings that we're attempting to mine, using a floating dredge and a coal-washing plant. I know more now about what I'm trying to do than I ever did before.

RAYMOND GARON

SURFACE MINER – LAKE INDUSTRIES

COAL MINERS

OF NOVA SCOTIA

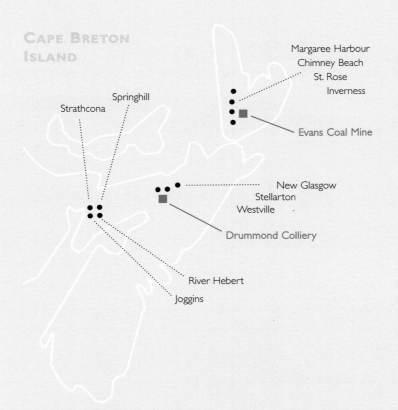

CAPE BRETON
ISLAND

Margaree Harbour
Chimney Beach
St. Rose
Inverness

Evans Coal Mine

Springhill

Strathcona

New Glasgow
Stellarton
Westville

Drummond Colliery

River Hebert

Joggins

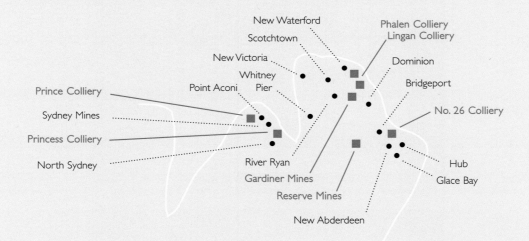

New Waterford

Scotchtown

New Victoria

Whitney
Pier

Point Aconi

Prince Colliery

Sydney Mines

Princess Colliery

North Sydney

River Ryan

Gardiner Mines

Reserve Mines

New Abberdeen

Phalen Colliery
Lingan Colliery

Dominion

Bridgeport

No. 26 Colliery

Hub

Glace Bay

NORTH SYDNEY / GLACE BAY AREA

● PHOTOGRAPH LOCATION   ■ COAL MINE LOCATION

BOB HACHEY

You ain't going to show this photo around too much, are you? I used to run around quite a bit and if somebody sees it, they might say, There's my old man. Actually, I've been married 35 years and have lived in this house 34 years.

I was 13 years old when I went into the Bayview Mine. I had no choice and neither did my brother. I joined the army, then came back and went to work in another small mine in Joggins. I worked there right up until 1961; then I went to River Hebert Mine and worked there until last year when it closed and I retired.

I worked 21 years for the Hillcrest Mining Company. When you add it up, I spent 44 years in the mines in the Joggins/River Hebert area. But if I'd had any brains, I would have stayed in the army.

You can't really describe what it was like, working underground on the thin pitching seams in the Joggins area. When you try to, people don't believe you.

ROMNEY McALONEY

My grandfather, my father and all my people were coal miners in this area back as far as I can remember. My grandfather opened up the Marsh Mine here. My father and his brother went in with him. They later sold out and worked in other mines in Joggins and River Hebert.

Just below here is the Standard Coal Mine. When I came of age, around 1939, there wasn't much work. After waiting a long time, I finally landed a job in the Standard Mine. Then I worked at the Riverside Mine about a mile from here until it closed in 1951. Starting in 1952, I worked at the River Hebert Mine until it closed in 1980.

I was underground manager at River Hebert for nine years. I was fortunate that management got after me to go try for papers. It's a privilege to handle so many men, to examine for gas and to fire shots. I was only 30 when I became underground manager. I was the youngest underground manager they ever had.

I could have had other work. In fact, for a couple of years, I managed a fried chicken store in Amherst. But I was glad to get back underground. I guess coal mining was in my blood. As far as working conditions go, it's hard, dirty work. You get black from the coal dust. Sometimes you're in water, sometimes you're in dust.

Miners and the public don't look at mining the same way. There are people who say, I wouldn't go down a coal mine, it's too dangerous. But if you look at the number of people killed in coal mines versus the number who are killed in car accidents, there's no comparison. It's safer to go down a coal mine to work than it is to get into your car and drive to Amherst. When we go down, we don't feel anything about danger. If we see a dangerous spot, we stay out of it or fix it up. A good deal of the fatal accidents in coal mining are from the men taking chances.

CONRAD ROBERTS

M y grandfather was a boss in the Victoria Mine. My dad was a lumberman and he worked in the woods all his life – I worked for a while with my father. I got into coal mining through some of my friends.

I started mining in 1954. We used to work during the winter then got laid off in summer because of falling coal sales. The Bayview Mine, where I was working, shut down in 1961 and they stripped her of equipment in 1962.

Dr. Cochrane opened up the River Hebert Mine. He was a doctor, a logger, a fireman and a mine operator. He showed up at the mine every day. River Hebert was a tough mine to work because you were on your knees all day and the seam pitched at 33 degrees. When I went into the mine, I used to shove boxes. Then I got my miner's papers.

In 1972, I got a permit to be a shot firer. Then I went to school in the winter of '72 and got my mine examiner's papers. Then, in 1979, I became underground manager working under a temporary permit. I'm now upgrading my studies because I hope to get a job in Springhill if they open a new mine. Yeah, if a new mine opened, that would take me to retirement.

I would go back to the coal mine today. Well, that's pretty well all you know when you work in a coal mine all your life. It's hard to start something different.

I lived in River Hebert most of my life. Then, 12 years ago, I bought this house in Joggins. I have three boys and one girl. I am the captain of the Joggins Fire Department. I keep active, I'm into everything. I am a trustee of the church, VP of the community association. I am a Draegerman.

JUDE MELANSON & WALTER LeBLANC

M y father worked 56 years in the mine. To tell the truth, we're supposed to be French but we originally came from Scotland. My dad could speak French and so could my brothers. Our family history goes back about 300 years in this area. Yeah, there used to be a large French community in River Hebert.

There used to be a lot of bootleggers around – every second house you'd find another one. My father helped drive the slope at the River Hebert Mine. He worked seven days a week and it took over a year to get the slope down through the mud, just to get to the coal. Driving that slope was difficult because of the mud. You'd take one shovel and two would come back in.

My last job was at the River Hebert Mine. I was there when they closed it down. I worked 31 years in the mines: seven years underground, then I went outside and went on trucking. My job was to haul coal from the mine to the Macaan Power Plant. Every day, I hauled about 150 tons.

The River Hebert Mine was very deep. Once, we found a tree underground that weighed 1,700 pounds. It went to the University. We used to find lots of fossils, mainly parts of trees. Some of them were quite beautiful. Some of the trees were preserved as well as when they were growing – you could count the rings. – *Jude Melanson*

RONALD BEATON

I started working in the coal mines in 1928, after I graduated from high school. In those days, there wasn't much money around for a college education and the mines were the only thing going. We had a large family with four brothers and three of them absolutely wouldn't have anything to do with the coal mines.

My dad was an official with one of the mines. When I started, I had a vague idea of what it was like to mine coal. I used to listen to a lot of the talk and in that way I used to visualize what I would see when I was in the mine.

I worked in No. 6 Mine, No. 7 Mine, No. 4 Mine. Then I moved to No. 1 Seam. Then I spent some time in the airforce. When I came back, I was moved to No. 2 Mine where I worked my way up to underground manager. I left the employment of the Cumberland Railway and Coal Company in 1957 and became inspector of mines for the mainland. The desk job didn't suit me too well.

Around that time, this small mine at River Hebert was about to go under and they persuaded me to take it over. I spent the next 21 years of my life keeping that mine going. I was the chief shareholder in the River Hebert Coal Company.

You must give the men at River Hebert credit because they were able to work hard with the old, obsolete machinery we had. They were all very co-operative in keeping that mine going as long as it did.

I found that if you matched men up who were congenial and equal in their skills, one would likely turn out to be the leader of the other. He would be the one to say, We do this now. Each would have his own task to do and that way, they'd get more coal out. Well, when you're on contract, the more coal you get out, the higher your pay is. So that's how most of these guys worked it: they became buddies.

I started coal mining April 9, 1928. I quit school and my father said, "If you're going to quit school, you'll have to come work with me in the coal mines." So that's what I did.

I started in No. 2 Mine and worked there until 1935. Then I went up into No. 1 Mine. It was an old seam of coal that was all broken. I worked in No. 1 Mine until 1954 and then they transferred all the No. 1 miners back into the No. 2 Mine.

No. 1 Mine was worked out and they had us stop working it because the seam was above the No. 2 seam and it was causing some severe bumps on the main slope of No. 2. The company thought there was a risk the main slope could be blocked and would trap some coal miners.

I worked there until just a month before the disaster occurred in 1958. Being chairman of the mine committee over a number of years, I knew this disaster was going to happen. I used to preach that the conditions were right for a major bump and that it would kill a lot of men. So I decided not to go down into No. 2 Mine no more.

As I said, I didn't work for a month. But the day of the disaster, in the morning, a very good friend of mine came and told me the three hand-loaded longwalls on three levels were lined up. I said, They're going to kill everybody in the mine. He said the officials were telling the men that the mine would never bump again. That night at 8:00 the disaster happened and it killed 79 men. My two brothers, aged 36 and 34, were killed as were my brother-in-law John Jackson and my first cousin, Tommy Tabor. I took part in the rescue and helped them get the bodies. — Joe E. Tabor

I started working in 1926 at the age of 17. Times were hard so I had to go to work during the summertime. I wanted to be a dentist — that's what I wanted to be — but I started in the mines at $2.75 a day.

I started out at No. 6 Mine at Junction Road, part of the Springhill coalfield owned by DOSCO. I started carrying timber up the chutes. That was a steeply pitched coal seam. In fact, it was so steep down at the bottom that they stopped mining. At various times, I worked No. 4 Mine, No. 2 Mine and No. 1 Mine.

After the severe bump in No. 2 Mine, I quit and went to the Annapolis Valley to work in construction. Then I came back and got a job at the Syndicate Mine in Springhill. That's where I finished. I was 60 when I quit and took a pension. Altogether, I had 41 years in the mines.

Now I look after the Hall here in Springhill. It's for the pensioners. They come here to play cards and there's a library upstairs. There's a little room back there where they can have a bottle of beer if they want. It keeps them off the streets.

Next year, we're going to Hawaii. My wife, she brought up seven in our family. She deserves the holiday — she's been a good woman. We've been married 44 years.

When you work with men for 25 or 30 years, they get to be like your brothers. You'd go down there and joke and carry on. If you got mad, you were beat. I remember when I started out at the bottom, a fellow sent me to get a "rib straightener." Except there's no such thing as a rib straightener. Of course, I was green. Yeah, just like one big family.

Unfortunately, I had to carry some of my friends out of the mines. I just happened to be on the opposite shift both times we had bad bumps here. Because I was an official, I had to go down immediately after it happened. — John Laurie

JOE E. TABOR & **JOHN** LAURIE

I was born in 1908, in Sunny Brae, about 15 miles from here. I suppose I was 18 when I started in the mine. Before then, I worked in the local bank. That's one thing I was good at – figures – but the banks weren't paying nothing. The wages in the mine were three times as high as what the bank was paying. I started out under the bankhead. It was a lot bigger operation than it is now.

There used to be three of us who worked under the bankhead. We had to look after the mine cars. Then I went firing the boilers. In the early days, everything was run by steam. Every day, we burned about 50 tons of coal. At one time, there must have been 20 men employed in the boiler room. We had one stoker boiler and the rest were all hand-fired.

These days, I'm running the mine for my brother's widow. My brother, Henry R. Thompson, recently died and he was the owner of the Drummond Coal Company. Henry worked in the powerhouse when I started mining. But he left because he had a talent for the stock market. On figures and machinery, he was hard to beat. He even taught engineering at night school. Later, he ended up owning this mine.

Oh yes, I enjoyed working in coal mining – I'd be lost without it. I had my chances to leave but I didn't. When the day was done, that was it. You didn't have to worry about nothing. Now that I'm in the office, I have to watch everything.

As manager, you've got a certain amount of responsibility all the time. If anything happens, you have to be right there. Counting myself, there are 42 men at the Drummond. I've put in 45 years at this mine. That's a long time to be working at one place. – *Chester Thompson*

I'll tell you, when you work with a thing 36 years, you get so that you don't think of nothing else. You're there all the time. It's more or less your life. You're not staying in it for the money you're getting. Let's put it this way: I think more than anything, you enjoy working with men.

I was born here in Westville. My grandfather worked at the Drummond Mine 40 years and my father worked here about 58 years. When he started here, he was 11 years old and there were 900 men. When I started, there were 400 men here. Now, we have only 30 men and three of them are my sons.

The Drummond Colliery is one of the oldest and longest-operating coal mines in Canada. There used to be a clay mine in association with the coal, and they made some of the best fire brick in the world.

It was 1946 when I came to Drummond Mine. I started off slewing boxes for my father and another man. Then I dug coal for 15 years. When the mine manager died, I wrote the exam and became the manager. It's a hard life but, if you're suited to it, you enjoy it. I think, more than anything, I enjoy working with the men. They don't bother anybody and they know their work.

In a busy job, boy, you're not only managing but you have to look after the pumps, the electricity and all the machinery. There's no spare time – brother, I'm telling you! It's much like anything else: sometimes you get disgusted, especially if you get too much stuff breaking at once. I haven't taken two weeks off, I think, in 15 years. – *George Dooley*

CHESTER THOMPSON, MINE OPERATOR & **GEORGE** DOOLEY, MINE MANAGER

TOMMY MacKAY, COAL MINER, GRANDFATHER; **FRED** CHAPMAN, RETIRED COAL MINER
& **MIKE** MacKAY, COAL MINER

DRUMMOND COLLIERY

DOUG PATTERSON & **WILLIE** EMBREE

ART HALE, **JOHNNY** COULET, **RANDY** DOOLEY & **GAR** LANGILLE

OZZIE SINCLAIR, **CYRIL** MYERS, **PETER** JOHNSTONE, **JAY** DOOLEY,
RICHARD LYONS & **EUGENE** JOHNSON

After school, I went to work in Halifax and I didn't like the city life. So my friend Huck got me a job at the Drummond. I was 18 when I started.

Before I first went underground, I was told what it was like but I couldn't imagine it being like that. The slope is the dangerous part. The roof of the slope is so low that you have to lean back in the cars. I couldn't get over that idea — men wanting to go up and down that slope every day. I said, "Whatever makes a guy want to do that?" But once you get out of the slope, it's a piece of cake.

I've had every job in the Drummond Mine — you name it and I've done it. And I have worked continuously since I started. I've even worked part of my summer vacation so I could take a week off in November to go hunting.

I've never been in another mine except for this one. I wouldn't know what to do in a mechanized mine, but I'd love to see one. I'm more than willing to learn something new. You know, you should get a picture of us with our shovels — that's all we got in this day and age, no mechanical loaders. All the mine cars at the Drummond are loaded by hand shovel: three boxes to the ton. The coal is a lot heavier than it used to be 'cause we're close to faults and the coal has more ash in it. But we carry on all day, laughing and making jokes and making the day go good.

I know there is nothing here. But there's talk of a new mine. After I got my shot-firer's papers, my friend said I could get a half-decent job at Grande Cache. But I said, "No, the money may be good but it's not my type of living."

My father and both grandfathers were coal miners. My father hurt his back in the mine and the doctor told him he'd never work again. Me, I don't know why I do it half the time. I guess it's bred in you or something. I've been told more than once that I was crazy for doing it, but I'd sooner be mining than anything else. My wife Donna says she tries not to think about me being a coal miner. She says if she thought about it all day, it would drive her crazy.

The Drummond Mine closed the year after this interview was conducted. Eugene Johnson and Mike MacKay (pg. 230) were two of the twenty-six coal miners who died on May 9, 1992 in the Westray Coal Mine not far from the Drummond Mine in Westville.

We are the Westville miners so tall and so proud
I'll say it again and we'll say it out loud
We are the Westville miners so tall and so proud
And that's what I think and I'll tell you right now

— *Eugene Johnson*

EUGENE JOHNSON

JIMMY JOHNSON

My mother said, "I don't know what we're going to do." At the time, I couldn't get a job nowhere. So there was no other place to go but the coal mines. She said, "I don't know what we're going to do, we ain't got nothing to eat."

I went down to the Alan Shaft Mine and got a job. On Monday afternoon – that was the 27th of January, 1930 – I went to work. I went from company work to working on the shaft bottom to running chains. I went from there to digging coal and I moved from there onto shot firing.

Don't you say a word – hard work! I worked in one place – we had 23 inches of coal, 23 inches high. You had to lay down to take a drink of water. You had to lay down to eat. Ah, boys, she was something. That's not much coal.

Ah, boys, don't you say a word, don't you say a word. There were a lot of men who died down there and a lot of horses killed. They had explosions every few years. I was in the explosion in 1950 – I was nearly killed. After the explosion, they gave me a watchman's job at the Alan Shaft. I was 22 years underground there. In 1961, I started working in the washplant which was far worse than working underground – too dusty.

I've been in the mines 52 years and I retire this fall. We had some bad conditions in the pit but in coal mining, you expect that. I enjoyed it because, I'll tell you, when we worked underground, we worked under good men.

ERNIE WRIGHT

Stellarton was a coal mining and railroading town. Years ago, a railroader would turn his nose up at a coal miner. They weren't supposed to walk on the same side of the street as a railroader. Coal mining was definitely looked down upon.

But a coal miner thinks he is as good – if not better – than any person in the world. There's no question about that.

My father was a coal miner and his family was from England. He was involved 48 years in the mines. He was killed in the Allan Shaft Mine – he was gassed. That was in 1932.

I started coal mining when I was 16 years old, in 1926. By 1933, I had worked up to a job as shot-firer. Then, in 1947, I became a manager at the Allan Shaft Mine. The Allan Shaft caught fire in 1950. We battled that fire for over a year. When it burned through a fire-stopping, there was only one thing left to do and that was to seal off the mine.

Not to flood it but to let the CH_4 build up to such an extent that it would put the fire out. After that was done, the company decided it wasn't practical to operate the mine any longer. I was

transferred to No. 20 Colliery. I left there in 1956 and went to work at the Beaver Mine. I helped open a mine called the Atlantic and I managed it. After leaving that mine, I went to Springhill where we had a nice operation at the Syndicate Mine. I wasn't too fussy about Cape Breton but I enjoyed working at Springhill. We had a good crowd of men there.

There were only two mines that I know made money. They were the Syndicate in Springhill and No. 7 Mine in Stellarton. Years ago, the old-timers made money but they robbed the mines.

You couldn't stay at mining 45 years like I did unless you liked it. Mining has been good to me but it's been tough, too. Don't ever get the idea that it's not. I have an excellent wife and, between her and my ambition, we made a good living out of it. My ambition at the time was to go as far in coal mining as I possibly could. Today there is one bitter spot: there are no pensions. It makes me boil.

I worked 25 years at the coal face, mainly here in Inverness. I was about 14-something when I started mining. My father worked on surface at the mine. At one time, there were 800 men working in No. 1 Mine, with 100 on surface. And there's coal there yet for thousands of years that they never took out. Both No. 1 and No. 5 coal was good but they claim there was a lot of sulphur. You couldn't bank that coal or she'd go on fire.

I played a lot of ball all my life and I did a lot of mining. Working in the mines and playing ball after work. I got a chance to play professional ball in 1920 when Glace Bay, Dominion and New Waterford were professional and they had some of the big-league fellows playing.

I was 18 and playing ball up here with a bunch from Glace Bay. This big fellow hit the ball and there was a fence between the field we were playing and MacDonald's farm. There were three men on the bases and two out. This fellow hit the ball a mile. I looked at it and it kept on going, going. When I got to the fence, I jumped it and caught the ball. They all said to me, "Why are you staying here, we've never seen the likes of that." Next, two fellows came down and offered me one hundred and twenty-five dollars a week and a job. I told them I couldn't go. They came after me twice.

One time, me and my brother loaded 38 tons by quarter to one and we pushed them boxes 500 feet before we were done. When I got home, my wife said, "The Sydney team is here, put on your uniform." I should have been in bed. I was so damn tired, I played third base. The ball came right through my legs.

The coach called me over and asked what was wrong. He had a quart of rum and he gave me a couple of big slugs. After that, I never missed another ball and I hit a couple of home runs to boot. The last one I hit, I jumped up and walked on my hands. I walked to home plate on my hands and they were all cheering. We used to do all those queer things.

I spent about 39 years in coal mining. I lost my arm in the mine. Everything came down and I got caught – I was the only one. I lost my arm and I got fifty-seven dollars a month for two years. I also had four ribs completely smashed. They couldn't figure out how I was able to live. After that accident, I couldn't even carry a bucket of coal to throw on the fire. But of course a coal mine is a lot safer today than it was then.

RETIRED COAL MINER

HUEY D. McISAAC

LYBISON MacKAY & SONS

My father worked 50 years in the Inverness Mines. I used to hear him talking with his friends about mining and I kind of graduated into coal mining myself. He took me underground when I was just seven. It was quite an experience. Yes, I think that had quite a bit to do with it.

I was born in Inverness. After the war, in 1948, I started at the old No. 1 Seam at Inverness. We were stealing pillars because the mine was winding down. I worked there six years.

I came to Chimney Corner the day after I married in 1951. I've been here ever since — about 30 years. So far, I have about 35 years in mining. We have eight children.

I started on the bankhead, dumping boxes. I think the Evans Coal Mine opened here in 1949, and it wasn't too long before I went underground. I'm still underground. I was underground manager for 20 years. I got my papers in the '50s.

I'm not retired but I have wound down. I'm not underground manager now. I do a little pumping, a little examining and a little shot-firing. That's all I want to do and I'm happy. I'm 60 years old and I'm talking another four or five years, but you never know. I may go beyond that. – *Lybison MacKay*

DEAN EVANS

My grandfather came in 1860 from Birmingham, England and he opened a coal mine here at Chimney Beach. He built two breakwaters here and had sailing boats to ship the coal around the Maritimes. The miners used steam power because there was no electricity in those days.

I was born in 1909, right in this house. My father operated his own mine in the '20s and '30s in this area. I worked with him in the early '40s and studied mining from the bottom up, right at the coalface, and got all my papers through the Province of Nova Scotia. In 1942, when my father closed the mine, I went to work for a government-operated mine in Inverness.

I worked for a small independent company whose owner couldn't read or write, but he could make a living. I said to myself, "I think I can do better." So I started a small mine in the middle of the woods where we'd located the outcrop of a seam my father had worked. I operated there until 1947 when we got power.

In 1950, we did some prospecting and found the outcrop of the main seam under four feet of overburden. We had been searching for that outcrop for years and years. In 1959, we got permission to abandon the old mine and we expanded the Evans Coal Mine at St. Rose. The mine was at its peak in the early '60s at about 50,000 tons a year.

In the '70s, things got tough. The cost of materials and labour went up and we weren't getting enough money from the coal. My boys are operating the mine now. To tell you the truth, I was happy there. I was doing things I was interested in and making money and providing jobs for 50 men. I knew all my men personally. That was one of the good aspects of being involved in the business: I knew the men and I knew their families.

I have a hobby, you know: restoring old cars such as 1930 Model-A Deluxe Roadsters and Model-T Fords.

ERNEST MacDONALD, GEORGE DEVEREAUX & PETER AU COIN

HOWARD MacKAY

I was born near St. Rose in 1893. I suppose I started coal mining in 1919 with Dean Evans' father. The mine was called the Eight-Foot Seam but it wasn't. It was a six-foot seam.

For the first ten years, I had charge of the mine myself, even though I didn't have underground manager's paper – I did have shot-firer's paper, though. I was in charge of the underground and Dean was in charge overall. Dean is my first cousin. We got along good and we never had a word. Although, when I went to work, there were some people around who said, "You won't work long for Dean."

I was no more than 12 years old when I first started working. I worked for Dean's father in the mine and I worked in his sawmill. I used to fire the boilers in the sawmill for 50 cents a day, 10 hours a day. I worked in the lumber woods in Maine as head chopper. I went to Calgary for six years. I farmed and I worked for the CNR.

I got married in 1922 and we had six children. In the summer, I fished lobsters and salmon and mackerel. I was keeping a boat, too. By working in the mine in the wintertime, I was going into debt because there wasn't enough work. I was fishing but I didn't enjoy it. I didn't like the water. I liked working in the mine by comparison. The mining was good. We used to have days off now and then, and take a little drink and have a good time around.

My wife was born in this house. Next year, we will have been married 60 years. You know, I say to myself, "That's too long to live with one woman." Oh, but we get along very good together. I can't complain. Oh yeah, I had lots of girlfriends before I got married.

HUGH COSTELLO

In 1938, I was on a runaway trip that killed 21 men. Oh yeah, I jumped off. I got banged around. There were two fellows killed right beside me when they jumped — the trailing rope killed them. I was lucky, that's all.

One fellow who lived, it was his second day in the pit. He went down on the trip for about two miles before he jumped. He wasn't even hurt. Afterwards, he asked me, "Does it go down that fast every morning?" That's the God's truth.

My father was an overman and a check weighman for years. He came from Scotland and was over 75 years of age when he quit. My father, God rest him, started out in the coal mines in Pennsylvania. I started when I was 16, in the Princess Colliery at Sydney Mines. I started on the pumps in 1927 and I operated and repaired them for at least 20 years.

I've lived here for about 50 years. Originally, my house was built as a company house. At one point, I started my own pit in the backyard, a crop pit. I had to go down a ladder 65 feet to get to the coal. You weren't supposed to do it but I had a barn building over the shaft. I kept going until I got the coal. The Chief of Police came and tried to shut me down but I said I had it all loaded to fire a shot. When I fired the shot, it blew out some boards from the building and they almost hit the Police Chief. And he started running.

A fall of stone cut off my foot in the coal mine — right off at the ankle. I now have an artificial leg. I was married only five months when I lost me foot. But I've worked 43 years since I lost me foot.

When I first started, there were mice in the mine. Then the rats came and the mice disappeared. I saw a big fellow laying down and a rat ran right over him. He said, "They don't bother you if you don't bother them."

ALBERT CLARK

M y father worked in the mine, I can't tell you how many years. During the 1925 strike, he got out of the mine and went into business. Then he went back into the mine for a short while before he died. He didn't like to see us going in the mine but there was nothing else around here at the time. I wore out about four pairs of shoes going to see the manager. He had an office with a window that had a shutter. When you went to see him, he would open the shutter to see what you wanted.

I worked in Florence Colliery (No. 3) for 22 years and I worked in the Princess for 21 years. Then I went on pension in 1969. I was 17 when I started. I did practically everything: I shifted pans, I loaded coal, I filled in for brushers, I trapped doors. Before I retired, I ran a donkey.

If I told you what some of the men had in their lunchpails during the Hungry '30s, you wouldn't believe me. Mickey Higgins, he was secretary-treasurer of the union, he was in the Princess one time, looking at the condition of the roof. He told me he saw people who went to bed after a supper of bread and turnips. That's what happens when a miner works only one, two or three days a week.

Coal mining was good. If you could get top management who were sincere and didn't take all the cream and leave you the skim milk – then coal mining was good. The worst thing about it was the coal dust. That was unhealthy and, in those days, there were no masks to wear. When they finally got masks, it was too late for us old people.

But you'd never meet a crowd like the coal miners. New fellas who might have heard me and another old-timer scrapping in the pit would say, "I wouldn't like to see those guys drinking on Saturday night, it'd be awful." But then another fella would say, "You should see them in the parlour – they're bosom buddies."

COAL MINING PENSIONERS

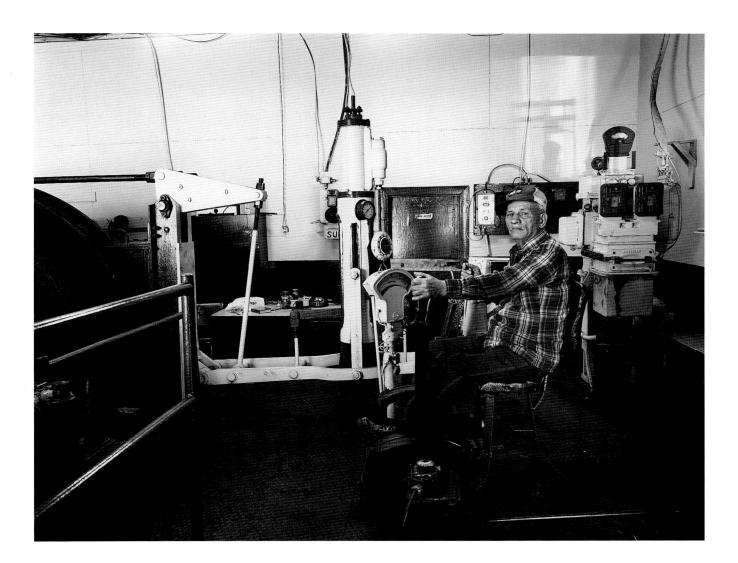

EDGAR BONNAR

I've worked 57 years. I started in 1925, when I was 16. I worked all my life in the Princess Colliery. My father was here as a blacksmith. My brother was here, all the family worked here. We are Scottish but I would say there was French in our name. I didn't work at the coal face, I started out on the tally. Then I went on the tipple/shaker where I worked on the picking table.

Everything was here on surface such as the rake and hoist engines. At first, they were all steam-operated, although now the hoist is electric. In those days, we had a big boiler room with five big boilers in it. I went to night school and got my stationary engineer's paper so I could operate the steam engines.

I became hoistman in 1945. I've worked on three different hoists. Unfortunately, I didn't get to meet a lot of men when I was the hoistman. No, it isn't one of the higher-paying jobs in the pit.

Now that the Princess Colliery is only open for underground tours for tourists, they aren't paying very much. Sure, I could retire, but I say you might as well keep going as long as you feel all right and your mind is okay. We had 40,000 tourists in the mine over the past two years. Some days, I've hoisted as many as 500 tourists down into the mine.

ARCHIE MacDONALD

My grandfather came here from Scotland – he was born there in 1834. In those days, all you needed to make a living was a piece of land close to the water. When he came here, he never had it so good. For later generations, it was easier to work in the coal mine or the steel plant than work those long, hard hours on the farm. I was among that generation.

I was brought up on a small farm near the No. 4 (Scotia Colliery). Farming didn't seem to have any future so in 1918 when I was 14, I started in the mine. The Scotia was a stone mine: the seam was thin and it was a poor place to work. When it closed, we were transferred to the Florence Colliery.

It was a big day when I went to the mine. It was the first job in which I made money. In other words, it meant I had my foot in the threshold of being a man – I would be somebody. I would be something on my own. A few years ago, I took a walk down by the old Scotia pit. I stood and looked down the road and in

my imagination, I could see the trips running and I could see the miners coming and going. It was nostalgia. It left me with a twinge of regret, you know.

In Florence, the colliery officials had steady work. These are the people who went to night school to get qualifications to hold some of those jobs. So, in 1925, I went to night school to get my overman's papers. I was 22. By 1930, I had my mine manager's certificate.

In 1939, I started as underground manager at the Princess Colliery. In 1948, I was appointed manager of the Princess and I was there until I retired in 1965. After I retired, I went back to school and obtained my high school grades 11 and 12. Then I started to work on a degree at the college. But they asked me to teach young mining supervisors at the College of Cape Breton, instead.

Now I'm looking after the Princess Colliery as a visitor centre for tourists.

SHAFT HEADFRAME

VINCE MacDONALD, VENTILATION SUPERVISOR; **REG** McINTYRE, MINE MANAGER

& **CYRIL** LeBLANC, PLANNING MANAGER

GERALD TOBIN & **JOHN** LOCKMAN

COAL MINERS — PRINCE COLLIERY

ANGUS MacINTYRE, UNDERGROUND MECHANIC; **JOHN** McMULLIN, COAL MINER
& **GREG** JOBES, COAL MINER

DICK BAXTER, **CLARENCE** PIKE, **DAVE** MATTHEWS, **LLOYD** TILLARD, **CARL** GOUTHRO & **RALPH** BOUTLIER

EDISON HOWATSON & **DENNIS** HARRIETHA

NICHOLAS GOYORFI & ABBIE NEVILLE

was born right here, in 1917, and I'm living here still. My dad was a foreman for the coal company and his dad was a foreman from the day it started, and both were involved in shipping coal. From the day the pier started, my family was involved in shipping.

I used to love the ships. My father took me on board a ship when I was six days old. I got in the habit of knowing each ship that came in. The ships would come in from all over the world – that used to make it so interesting.

In 1930, the "black pan" kept a lot of us kids alive. You ever hear tell of the black pan? All the kids on the street would go down to fish and go aboard this boat at 8:00 at night. Whatever was left from supper would all be in a pot. We'd eat from the black pan with the sailors who were going off watch at 8:00 p.m.

I worked on the pier for 47 years. My first job was tallying coal – that is, checking all the cars that were coming and going. When the war started, I went with the superintendent as his clerk because all the convoys were coming in to the bunker. Sometimes there would be as many as 150 ships in Sydney Harbour at a time. You had to get up every morning at 4:00 and call the Navy dockyards and find out what ships had come in, where they were berthing and whether they wanted a bunker. Then I would start trying to bring them in, one at a time. I'd be there from four in the morning until 10 at night with no pay.

The coal company owned the Black Diamond Fleet. During 1925, coal was shipped from Sydney to England because they had a big strike on there. We also supplied coal for the steam locomotives on the railways. In 1954, I became shipping superintendent. Everything was under my control. On the pier, foreigners and coloured people were not allowed to work – that is, until I took over. – *Abbie Neville*

I was born up at Low Point in 1902. I couldn't really tell you when my family came to Cape Breton from Scotland. It was way, way back. I started mining January 9, 1924 and I worked to March 1967. I spent 52 years, one month, 17 days and four hours underground. I worked mainly on the coal face, but I also worked in construction, roof bolting and diamond drilling.

On Wednesday, July 25, 1917, I turned 15. It was a nice sunny morning, all cheerful. We all went to work. I was a brakeholder at the time. That morning, we were waiting in our place for some boxes to come down on the cage. Just then, a fellow passed by – he was going for a light.

It was all oil pit lamps then. You sometimes had to walk a half-mile to get a new lamp. Everybody who had a dark lamp would give it to you to take and replace. You might have five or six lamps by the time you got past.

Anyhow, just after the guy passed us, there was a loud shot and we were pelted against the wall. There was fire with it and my face was singed. I realized something serious had happened, but I didn't know it was an explosion. The force put out my light and there was dust around. So I followed the rope down to the bottom. I guess if I had panicked, I would have died.

There were 65 killed in that explosion. It was a good thing I didn't have a light because when I went down to the bottom, I would have seen all those fellows killed. I don't know what I would have done.

I never took part in too much storytelling. I'll tell you the simple reason why: when a bunch of miners got together, jumping Christ, it was always women, women. But there were humorous things happening every day. We were blessed with this one guy in the mine here, No. 18 Colliery, who was a great storyteller. He could make a story real and he had a new one every day. He didn't smoke or drink but he died suddenly with a tumour on the brain. – *Toots MacNeil*

U ntil this latest machinery was introduced, it was a pleasure to go to work. You got up in the morning and had a bite of breakfast, you knew you were going to have a good day. When the pan line stopped along the wall, you'd all be together, telling jokes and carrying on.

The first modern equipment was brought into No. 18 Colliery in New Victoria. When you had the new equipment, you had to work harder and longer hours.

I started in 1935 in Reserve, coupling boxes, then I started on the coal face. You had to wear knee pads and you quickly wore out the toe on your boots from crawling around.

In 1947, during the big strike, my buddy tried to talk me into going out west. But I didn't and always regretted it because he got a good job. I ended up bootlegging coal here at a bootleg pit.

I thought the world of coal mining. I was on a farm in Pictou County for 13 years when I was young. I left the farm and said, I'm going to Cape Breton where I belong – there's no money up here. I made the right decision to be a coal miner. I would go back to the mine again if I was younger and able to work. Sure, it was dirty, but there were lots of good times.

One time, we were breaking for lunch and our crew was all together. All of a sudden, I could hear this tick, tick, tick. It sounded like a bomb ready to go off. "What the Christ is that," I said to Johnny, "What the hell is that?" Joe opened his lunch can. He had an alarm clock in there, wrapped in a piece of cloth. Laugh me Holy. A big alarm clock because he had no watch. – *George T. McNeil*

TOOTS MacNEIL & **GEORGE T.** McNEIL

MINNIE BURKE

I can remember when I was four years old and my father, a coal miner, went away to the Springhill Mines.

My husband Joe didn't die in the pit. He took miner's asthma and I never got a single cent. He started mining when he was 14. I was glad when he got out of the pits – I don't like coal mining. I was 22 when I got married and I'm 75 now. As old as I am, I could have been married 15 times. That's the god's truth.

Joe was good to me. He wouldn't let me smoke. You just had to do what he said. We had ten children, six sons and four daughters. By the crying Moses, I would take a thousand kids tomorrow, if I was younger.

I don't like living in the seniors' lodge – you can't do nothing except play cards and tell stories. Oh my God, we used to sit up for hours, telling stories. I still go to the Heather Beverage Room every Wednesday night but the damn old drums deafen me – bloody old drums. I still play the harmonica at the Heather. I'd like to get a good mouth organ, one with double reeds. I gave my old one to my son. I used to play the accordion but I gave that up.

We all loved music. Oh, I used to sing good songs, old country songs. I knew lots of songs! Holy smoking Moses, I'd follow the piper to the devil.

WAYNE POWER & **CHARLIE** HELLE

GLENN MARSH, UNDERGROUND SUPERVISOR & **BRIAN** CAMPBELL, CHAIN RUNNER

LINGAN COLLIERY

WILFRED HURLEY, CHECK #3933 & **ANDY** HYNES, CHECK #4364

NO. 3 DEEP DEVELOPMENT - LINGAN COLLIERY

PLEMAN WOODLAND

GARY COVEY, MECHANIC & **JAY** DOOLEY, LONGWALL SUPERVISOR

LINGAN COLLIERY

BILLY BURCHELL

LINGAN – PHALEN 1987 MINE RESCUE TEAM

TERRY O'BRIEN, BRAKEMAN & **RANDY** GILMET, CONDUCTOR

JIMMY P. McNEIL

I'm 76 years old and still have a little on the ball. I was mayor of this town for 13 years and town councillor for 14 years. In 1953, when I first became mayor, the first roads were paved here in New Waterford. In those days, in the fall of the year, you had to wear hipboots to go out from one place to another. There was no sewer system – it was all outdoor toilets.

There were about 10,000 people in New Waterford in those days. There are about 950 people here now.

Don't worry, we knew about mining and, goddamnit, we worked hard. My shirt was wet every day. I hired on in 1917, the week of the No. 12 explosion. I worked in four different pits a total of 51 years. The explosion didn't stop me from working underground. I enjoyed every minute of it.

I was running the hoist engine and the fan engine at No. 12 Colliery return airway when I went on pension. I could have been doing that work until I was 100 – what a job!

We had some pleasure too – it wasn't all hardship. Like any job, it's only what you make of it. If you want to be a grouch and find fault with everything you see, there is nothing satisfactory to you. But if you see things the way you should see them and feel that the world don't owe you a living, you'll get along. You'll like it and people will like you.

It's not that a person didn't have any education. As far as education is concerned, I had grade 6 and I was mayor of this town. See, when I quit school, I went to night school and I got a lot of practical learning. I found out that the earth is round.

I practically built this house by hand. It's awful for a man to be living alone in a big house like this. Walk around and have a look at the place. I want you to see that the house is well-kept. I have a coal stove in the house.

GUS McNEIL

How I felt about mining: You'd wash up in the wash house and you'd hang up your clothes and you'd forget about the pit. I left Cape Breton to go out west during the strike in 1925/26. I went tramping as far west as I could, which was to Alberta. I came back to Cape Breton just in time for the riot at the Lake here when that fellow was killed. Then there was an uncle of mine wounded by a bullet in his stomach that had been fired by the police.

I took the train from Sydney, heading west. By the time we reached Fredericton, every window in the train was broke — Jesus, it used to be a rough bastardly train because everybody was goofed up. I went as far as Lethbridge where I got a job working on a thresher, driving horses and loading up the bundle racks. It was no easy job. I was about 19 years old at the time.

I always travelled alone. It was better that way when you were looking for work. Except that you were never alone in the box cars. Those were about the best years I spent.

After threshing and bridge construction, I got a job in a coal mine at Coalhurst near Kipp, Alberta. You had to walk about two miles from the mine to Coalhurst. It was a big mine with 450 men working and a 900-foot shaft. I went back to Coalhurst a few years ago: nothing there but the stone dump.

Every summer, I used to go on the road. That's how I saw much of this country.

MICKEY MULLENS

started in No. 14 Colliery in New Waterford, driving a horse. I was 15 years of age and I then did chain-running. My mother was anxious for me. She wanted me to work in a store or something but, at that time, the only place you could make money was in the mine.

In the Coal Mines Regulations Act, it is specified that a "miner" is anybody who works in or around the mines. To become a "coal miner," you have to get certification. The officials in charge cannot put you at the working face unless you have a certificate. I had the ambition to get to the coal face. I wrote for mine examiner, for shot-firer, then for overman. The next step was underground manager. I got my manager's papers. I was smart in school, particularly in math, and my main interest was to make money.

I was underground manager at No. 12 Colliery for two years. Then I was mine manager for 1,200 miners at No. 16 Colliery for 12 years. In dealing with people, you have to be

a psychologist. The key is proper communication. In 1973, when I was pensioned with the company, I was over 65 years of age. I wound up working 50 years.

I'll tell you why I stayed in coal mining: the two basic industries here in Cape Breton are coal and steel, and I wasn't fussy towards steel. With the coal company, the money was good and I was learning a wealth of information and the people I worked with were good people. I had no reason to leave. It gets in your blood. At one time, there were 14,000 men working in the mines on Cape Breton.

I've always been involved in teaching. I took on the job of co-ordinating night school on a part-time basis. After retiring in 1973, I was involved in the College of Cape Breton.

I'll tell you about an experience I had when I was a young fellow. I was working on the backshift in No. 16 Mine at the time. At the beginning of the shift, the men walked over and sat on the rake, waiting for it to take us down. After a while, as more men got on, it got heavier and heavier. The rake started to move slowly but when she got down a little piece, she was going pretty fast. Some of the guys started jumping off and they got broken legs, broken arms and all this kind of thing, it was going that fast. There was no engineer and the engine house was locked.

Then one fellow broke the lock of the engine house and stopped the rake just before it reached the end. If it had hit the end, we would all have been killed. A young fellow in front of me was trying to jump off at the low spots but I held on to him. I probably saved him from getting seriously hurt.

That's something about miners: if one of them is in danger, the others will do anything to get him out of danger. I worked 42 years in the mines. I worked in both No. 16 Colliery and No. 18 Colliery, mostly loading coal on coal production. Then I got on the machines, continuous miners.

I went overseas during the war. You'd think I would have learned something by being overseas, but I come right back and went in the mines.

I retired in 1966. I didn't want to go to the No. 12 Colliery so I said, That's it. I got nothing when I retired. All I get now is a ton of coal at a cheap rate. A miner's life is a hard life. I'll tell you what's good about mining, though: the fellowship, the comradeship. There's something about a miner that's different than anybody else.

In New Waterford, there's a tremendous amount of musical talent. I play the accordion. I play good and I play lousy and I will give you a little of both. — *Clem Mossop*

I drank a lot in my time. Good rum — straight — for three dollars a quart, tasty and good quality. Smuggled in from the Barbados. Strong enough to make a rabbit kick a bulldog.

I started in the coal mines in 1925, in No. 12 Colliery on the eleventh day of January. I spent about 32 years underground — 44 years of service altogether. When I went into coal mining, it was every young lad's ambition to get into the mines. That was the only thing here in this area. I had an old suit — dear God, there were 1,000 patches on it — and after my seventeenth visit to the mine office in my suit, I got hired.

My job for about 20 years was underground mine electrician. The last couple of years, I was on surface and in the mine whenever there was trouble. There's no man who's worked in mining more than 10 years who has not experienced injury of some kind. One time, I got horribly squeezed underground. That time, the Old Fellow with the hook reached out and missed by a whisker. When you're down there, you've got to think for yourself and the other guy with you.

The company, Dominion Coal, wanted all your sweat and some of your blood if they could get it. They never contributed to the town. It was always take out, take out.

I had seven children. One of my sons was working in No. 12 Colliery. He was doing one of the hardest jobs, brushing, and we were still receiving the baby bonus. The youngest boy is one of the vice-presidents of the coal company today. The other guy is president of the Labour and Trades Council. Joe, the oldest boy, is an electrical contractor.

I used to play the sax in a fife and drum band when I was a kid. Somebody was hauling an old piano to the dump and I bought it for five dollars and fixed it up. My daughter is a piano player in a band. I finally had to go on pension because I was crippled. I've had two spine operations. — *Raymond MacNeil*

CLEM MOSSOP, RETIRED COAL MINER & **RAYMOND** MACNEIL, RETIRED MINE ELECTRICIAN

DOMINIC NEMIS

A long time ago, I hauled six barrels of rum for a fellow. He gave me three barrels and he had three barrels for himself. I sold the barrels for six hundred dollars. I wouldn't sell it cheap, it went for the highest price. The mounted police stopped me with the rum. They said, "Here is a ticket for speeding." I had a 1935 Chevy. That son-of-a-whore could fly – I was going 85 miles an hour. A fifteen-dollar ticket. That's all I made that week in the pit – fifteen goddamn dollars.

I was born in 1908, in Italy. My father came from Italy with a ticket in his packsack. He couldn't read English. I started in No. 15 Colliery in 1923, driving a horse underground. The mine worked only two years then a strike shut her down in 1925. From there, I went to loading coal in No. 14 Colliery.

I worked 34 years in the pit. I worked for every cent I got. They'd never give me nothing for nothing. I'll tell you why I worked as long as I did: I had grade 4 and that was the place for me,

in the coal mines, because I didn't know any better. It was something to get a job in one of the mines. Now, the new mining, we wouldn't know anything about it. Don't worry, we knew about mining and we worked goddamn hard. My shirt was wet every day. Not like now – they've got machines producing coal now.

I had only grade 4 but I picked up plenty since then. I always read the paper and I know everything that's going on and I'm interested in politics. I was president of the Local 4526 UMW for 22 years.

One time, I was light heavyweight boxing champion. My brothers were fighters but I wasn't much of one. Everybody used to beat me, but one time I got in shape and I won the title. I knocked out – well, Jesus! – just like Joe Lewis. You fight anybody when you're in shape. You don't feel the punches, you don't feel nothing.

FREDDY Y. McNEIL

I've played the fiddle most of my life. These days, I play it over the CB radio late at night for my friends. I also sing old mining songs by Doc Watson and play along with my old guitar. You remember his song, "Daddy, Don't Go In The Mine Today"?

I worked 43 years underground and seven years on surface. By Christ, I had a shovel in my hands for 35 years. Until I retired – that was just recently – I was a hoistman along the main deep.

I've got silicosis. When I was operating the hoist, I used to have extra time on my hands so I would make lunch pails for the other miners. I used to charge three dollars a pail. About every third fellow said he'd pay you next week, but he'd never pay.

Oh, I used to find the materials for making the lunch pails. Sometimes I even had the officials help me find the materials. I used to make all kinds of things while I worked on the hoist. But after the explosion, they said that anybody who had an aluminum can had to do away with it.

You have to be a snotty bastard to be a manager. You have to have that killer instinct in you and you have to be on the rough side.

RANDEL McLAIN

JOHN KELLEY, OPERATIONS MANAGER; **GEORGE** WHITE, GENERAL MANAGER;
WILFRED DUCETTE, DEVELOPMENT SHIFT MANAGER & **JIM** MacLAUGHLIN, MINING ENGINEER

DONNY TRACY & **KEVIN** MORRISON

VICTOR KOZIL, **FRANCIS** CAMERON, **GEORGE** NICHOLSON, **JIMMY** COULLING,
JOE GILLET, **MARSHALL** PRINCE, **CRAWFORD** LLOYD & **ROBERT** BETTS

BOBBY KENNEDY, **RICKY** KELLEY, **JERRY** McNEIL & **SAKIE** CORMIER

ELECTRICIANS — PHALEN COLLIERY

DOUG SNOW, **SANDY** MacDONALD, **BILL** LUDLOW & **JOE** BYRNE

COAL MINERS – PHALEN COLLIERY

JOE MacDONALD, SURFACE SUPERVISOR &
DUNCAN MacKINNON, MAINTENANCE NO. 2 SLOPE BELT DRIVE

GERALD MacDONALD & **RON** SEYMOUR

UNDERGROUND CONSTRUCTION – PHALEN COLLIERY

DENISE AND CONDO BAGGIO

When my son left home to go to work with his father in the mine, I cried and cried all day. I never thought I would see him alive again. I said to Condo, "You keep him alongside of you," I said, "See that he don't get hurt or anything."

It's not too long ago that I found out that his first day in the mine, my son went one way and Condo went the other when they got underground. – *Denise Baggio*

I was born in 1903 in northern Italy. I arrived here from Italy on a Sunday and, three days later, I started work in the mines. I worked in coal for 20 years and then on stone for 28 years.

I retired in 1968 from No. 26 Colliery. Prior to that, I worked in No. 1A and No. 1B Mines. At one time, there used to be 150 Italian miners around here when No. 1B Mine was going full-steam. That was in 1925/26, after the big strike.

I have one son who has been working 22 years in No. 26 Colliery. He puts up arch supports. – *Condo Baggio*

ERIC SCOTT

started working for the Dominion Coal Company in the No. 1A Colliery here in Dominion – that was just after my 17th birthday. One day, I remember, I was loading the wooden hoppers and cleaning up under the bank with a long-handled shovel. Mr. Munroe, he was manager of the mine, he came over and said, "Eric, how old are you? I've been watching you from the office and you're doing a man's work." "Well, I said, I'm 17 years old."

I was too honest. I should have said I was 18 because at the time, you had to be 18 to get a man's wages. I ended up working 30 years in the mine – 18 years of those as a watchman.

Often when I hear the song "Sixteen Tons" – you remember that song? – I remember Murdock Sampson and I working together to load 16 tons. We cut the coal, bored it and shot it down. But to tell you the truth, I never liked the pit. I decided to get on the watchman's staff.

In the days I was watchman, it was all open – everything was open. Now they have gates, fences and little houses. There used to be fellows going around with trucks and carts, stealing coal and timber. Sometimes there was a family with little children and the father was dead. They used to come up on the railroad to pick coal. I'd never say anything to them. I'd always let them come and pick coal but if they were stealing it – that was a different thing. A lot of people would steal the coal and sell it.

I remember one day – it was on a Saturday after dinner – and the boss came up. We were in the manager's garage. He said, "Look at the young fellows down on the track, gathering up coal." I said, "Yes, their father is dead and they're a young family growing up." I said, "I always let them pick coal on Saturdays. He said, "You done right. You do right."

I boxed. The first time I was in a boxing ring was in 1918. I'll tell you, when I was 70 years old, I went over to the New Waterford one Sunday and I lifted 420 pounds. I lifted 160 pounds with my right arm. I believe I was the strongest man in the Dominion Coal Company.

MICKEY ELLSWORTH

I was born right next door in 1909. My father was a coal miner, a policeman and a blacksmith. He came from Low Point. I think I was 17 when I started with the Dominion Coal Company in Glace Bay. There was a fellow who worked with us who finally got hanged in the Sydney jail. He killed a taxi driver in Glace Bay. There was another fellow who worked with us at the same time — he should have got hanged but he was cleared.

I listen to the scanner to get police reports. I find out what's going on even before it happens. I got about 25 years of mining. I was running engine and then I went on the motor road. I was a dispatcher of the mine cars underground for about 15 years.

I had to quit mining in 1955 because I got smashed up. I got between a full box and a motor. The box tipped over and I was in between. It was my own stupidity — I blame myself more than anybody else.

But I'd sooner work in a mine than a grocery store because in a mine, you've got more freedom and everybody's buddy-buddy. You couldn't do nothing on Sunday but Monday morning everybody would know about it. And you were a fool if you said you didn't. There's an old bunch of them that used to get in one box going in the mine on the rake. Andrew H. and Bill M. and the rest, and they'd start singing hymns, oh my. Oh hell yes, it was nice! I play the linoleum, that's all.

Every morning at 10:00, as long as I'm able to move, I go up and have a beer. I'm usually home by 1:00. I never go out at night anymore because the way it is now, you don't trust who you meet on the street.

I like to have budgie birds. They make a little bit more noise than myself growling.

LLOYD WILSON & GRANDSON, **STEVE** GIBSON

RETIRED COAL MINER

Row STREET

SUBMARINE COAL MINE

JIMMY SMITH

ELLIS KELLAND

LEO O'NEIL

BILL TALBOTT

M y dad came here from New Glasgow in 1902. My parents had eight boys and three girls – and every damn one of us worked in the coal mine. Every damn one of us. I've seen eight of us in the coal mine at one time. I said, "Jeez, if we ever have an explosion down here, they're going to wipe us out."

Now we're whittled down to two boys and two girls. All the rest are dead. I had no education – when I left school, I was in grade 3. What bit of knowledge I have, I picked up myself. The way I figured it, if one man can learn to read and write, why can't another man learn to read and write?

I started to work in the pits when I was 13, trapping. Then I left the mine and went into the First World War. I was with the Cape Breton Highlanders. In 1940, I joined the army again. After the war, I came back to No. 2. In 1929, I got my overman's papers. In 1940, I wrote for underground manager but I failed.

This is what you call a coal company house. I got one of the nicest views in the Glace Bay area. Before my time, there used to be a coal mine right below us. During the 1925 strike, we'd go down there and dig coal.

All the pictures you see on the wall – some of these are my grandchildren and some are my daughters. We had nine children of our own and adopted three girls and a boy. People told me, "You're crazy." I said, "No, I'm not." I said, "If we're not rewarded down here, maybe God will reward us in Heaven." I don't know how we did it on a coal miner's small wages, but we brought them all up.

I was president of the Army-Navy Club four times. In the '50s, I was president of the old Hub Club. I've been a service officer of the Legion for about 30 years.

CLIFF WEATHERBEE, COAL MINER; BUSTER BEARCAT WEATHERBEE, RETIRED COAL MINER & HAROLD WEATHERBEE, COAL MINER

My father was a blacksmith for the coal company all his life. I don't doubt I like horses because of my father. He'd pull off the old shoes, make new ones and put them on again. He did 28 horses in one day, once.

I went into the mines when I was 19. I was a driver. I was working with a fellow who had a big horse by the name of Jim. We started in pulling eight boxes and another trip was coming out. The two horses crashed. Well, mister, that took the joy out of me. They had to kill one horse right there. "Oh, mister," I said, "No more of that driving for me."

So I worked my way up till I got a place on the coal face. She wasn't all sunshine, though. I've seen lots of times, you'd go down and wouldn't open your can until you went home. I quit in 1955. I was trying to work here in my yard as well as in the pit. I'd work all day, come back and get a bite to eat. Then crawl under

somebody's car and put a motor in or a transmission. Work there until 11:00 at night, then back into the house and into the pits again the next day.

But oh shit, yes! I enjoyed working more on cars than in the pit. Listen, it's a job, but if it was left to me for my boys, I would sooner see them go pick up horseshit than go into the mines. Too dangerous.

I always had ponies, working and jogging horses. I paid one dollar for my first racing horse. He was an old one. But I started training him and he came to think he was a three-year-old. That old son-of-a-whore, I must have put 25 bags of Royal Horse Feed and Oats in him — I just took to that old son-of-a-gun. First thing you know, we could beat all them horses up here on the track. That old horse made me a dollar and I had a lot of fun with him. — *Buster Bearcat Weatherbee*

D.J. O'BRIEN, RETIRED MINE SAMPLER, **HAROLD** ASHE, RETIRED COAL MINER

& **HAROLD** ASHE JR.

H AROLD: I just kind of sit in here and relax. I come out and light the old fire. My wife comes out once in a while and I'll bring the bottle and pour a nice drink.

I'm 73 so I figure I got 27 more years to live. After 100, you can do what you want with me. With that attitude, I get up with a smile every morning. I get up in the morning and thank the Lord that I have seen the daylight and that I still have the strength to carry on another day.

I started mining in 1934 and I got 35 years in. Then I took early retirement in 1969. I started in No. 20 Colliery and stayed there until I quit. My family came from Newfoundland. My father was a miner – he worked in No. 2. That No. 2 Colliery, located in New Aberdeen, was a wonderful mine. The seam was flat.

When I took early retirement, there were five of us at home and I was only getting one hundred and two dollars every two weeks.

I got two boys in the mines – they're down at Lingan and Phalen. I definitely would do it all over again. I tell my boys I'd like to go back underground. It did me heart good to get up at 4:00 in the morning and go to work.

HAROLD JR: I joined the army in 1958 and stayed on until 1981 – that's 23 years in the army. After that, I got a pension but not enough to live on. I'm not going in no mine! That's why I left here in the first place.

D.J. O'BRIEN: I came to Cape Breton from Newfoundland when I was two years old. I'll be 65 in two more years, then I'll tip my cap and get my gold watch. Altogether, I've been working 45 years in coal with Dominion Coal and the Department of Mines.

JASON MacDONALD & JIMMY BREWER

Did you ever have a chew of tobacco, the good stuff — MacDonald's? The sickest I ever got was from an Italian cigar, chewing a piece of it. The Italians used to chew them all the time in the pit — you know those queer twisters. You break a piece off and put it in. Oh my Jesus, I turned green and dizzy and, oh boy, Jesus, the sickest I ever got in my whole life.

I was about 15 when I started mining in 1941. I remember it was on a Saturday morning and I got a call to go down. From Dominion to No. 26 Colliery was about a mile and a half, and me and my friend took off down the railway track in a hurry to get there. We started work that night. It took no examination, no nothing.

On my first shift, Paddy the road maker sent me across the deep to get the "road bender." I didn't know what that was and when I asked a couple of miners where I could find the road bender, they went back and gave Paddy hell for sending me across the deep.

If you go up the shore where they used to dump over the bank, you can pick up a bucket of No. 26 coal. It's as good as three buckets of the other stuff. Lots of heat! That No. 26 coal: the further she goes down, the better she gets. — *Jimmy Brewer*

HARRY MURPHY

I started mining when I was 16 years old. I didn't retire – they retired me. When you're 65, you go. My father was a miner but he didn't get me started. I was anxious to get out of school. When I first went to work, it was like going to school again because my schoolfriends were also going to work for the first time. If you're big enough, you're old enough.

I started in No. 2 Mine and she became No. 20 when No. 2 closed out. Same shaft. When I retired in 1975, I was in No. 26, working as a motorman and doing road maintenance. I loaded coal and I brushed stone. I did road-digging and road-making.

When we retired, we didn't receive a watch or a clock. There were too many coal miners for them to do something like that. In this part of the country, working in the mines is the best-paying job around. Oh yes, I liked the mines but I didn't know anything else. My son started in the coal mines here. Then, when they started to phase out the mines here, he went into hard-rock mining.

BILLIE PITTMAN

I was born right in New Aberdeen. My grandfather came to Glace Bay in 1804, so I was from one of the pioneering families of the town. My father was a coal miner and he went overseas during the first war and got wounded three times. Then he came back and went coal mining again and he worked around 50 years.

I started work in the old No. 9 Colliery when I was 13 years of age. I worked there until the mine closed in 1922. Then I transferred to No. 2 Colliery which was considered the big producer at the time. And I worked there all my life, you know, until the war broke out. I happened to belong to the militia and, when the war broke out, I had to report. I went overseas with the medical unit and served there until the war was over.

Then I went back to No. 20 Mine. I worked the plugs and was in the explosion there. I went from trapping to flagging, driving, loading coal, being shot-firer and overman. I was overman for years. Then the last few years, I was on surface. I retired in 1960 because they were trying to close down the mines. And when you're 60, you're out. They credited me with 50 years of service in the mines. I didn't really have the 50 years but they added on my army service.

BURT GUTHRO

f I had to do it over again, I would. Miners are a great bunch of people to work with. When you got into the cage in the morning and looked around, you knew every damn one. You knew where they worked and where they lived.

When I first went to the No. 1B Mine, there were 70 horses underground. The one stable I had on the north side had 36 horses – that was a big stable. In the beginning, I didn't want to take that bloody job of head stableman. I was making more money at my previous job. Anyway, being stableman meant I got a seven-day-a-week job because the horses had to be fed and watered every day.

But I think I was born in a horse stall. I looked after horses all my life. I could tell you a lot of queer stories about horses. There was one white horse named Jumbo. He used to go around and open the men's lunch cans. When he found a lunch can, he'd put a big foot on it to flatten the can, then he'd have a snack.

The horse I had in my yard was an old pit pony. He was 27 years old when I finally got rid of him. At the end, he would lay down and lay there so long that he'd have trouble getting up.

Horseshoe Dan was in charge of one of the three underground stables at the No. 1B Mine. He used to talk to the horses and to himself. He was a queer old shit. He told me that when he first went to work in the Sterling Yard, the coal company had 700 horses.

The end of horses working underground was in 1965. The photograph, "The Last Trip," was taken in No. 26 Colliery. Somebody asked me to put a harness on Emma and hook her to a box of coal because they wanted to take a picture of her.

I worked 36 years in the pit. You see, I was an old boy when I started in the pit, because previously I'd been a dairy farmer and worked with horses on the farm.

DANNY D.J. O'BRIEN

The mine did have a magnetic draw to it. We used to say that if you came from the mine and then took another job somewhere else, it'd be like going to church with no clothes on.

But I always thought I wanted to do other things.

I ended up underground. I was working on the spillage. I was loading away and the sweat was running down my face and back. A guy came by in coveralls, looking important. When I asked what he did, they said he was a surveyor. I said, "That's the job for me." "No, they said, don't be talking so foolish. That's a job for the manager's son or the superintendent's nephew." I said, "And for Mick O'Brien's son."

At this point, I got into night school. Of course, I worked every day and would even take an extra shift — oh yeah, I was greedy for that. So, by golly, this one day I never showed up for work. By Christ, they were saying, O'Brien must be sick. And when I didn't show up the second day, Christ, he must be dead because he'd never miss a shift.

What I was doing was getting interviewed to get into the engineering department on the survey staff. I was accepted. I worked for the company from 1941 to 1965. I did surveying for nine years. Then I went into the coal sampling and became assistant quality control manager. We sampled the stone dust and coal. Then, in 1965, I joined the Federal Department of Mines where I continued to take samples on a regular basis from all the Nova Scotia mines and others in Canada.

There's a story about a guy who loaded 10 boxes a shift. He always tried to get 10 boxes, no matter what. His friend would take his time and load six boxes. That's all he wanted — that was enough. By God, the fellow who loaded 10, he died. His friend went to see him at the graveside. He said, "George, once the score was 10 to 6. Now she's 6 to 0."

RONALD MacKINNON

I was born in 1904 near Sydney. In 1921, I dropped out of school and got a job with the company, carrying mail. I moved on to become office boy and blueprint boy, working around the tables and cleaning up. Then I gradually started working on the tables, drawing and doing odd drafting jobs. In 1932, the company moved our office from Glace Bay to Sydney. Working on the tables meant plotting up the notes from the underground surveyors who sent in their notes each day.

Each day, we used to plot on tracings, bringing them up to date. Once a month, we would plot the rooms. The rooms were all surveyed in. The general program was very accurate: the basic plan was one inch to 200 feet. It was our job to give the bearings to the surveyor to lay out the crosscuts and the rooms. The miners generally followed the lines laid out by the surveyors.

After working on the tables for a long while, I got mixed up with supervision of the underground and surface surveyors. I never did actual surveying even though I became the chief surveyor in the '40s. I would occasionally go down in the pit and show the surveyors how to set up for tricky jobs. I got very involved in mine planning for the Dominion Coal Company. I looked after 11 mines in Cape Breton, plus the mines in the Springhill area. I never produced any coal but I was responsible for a lot of coal being produced.

I retired the month I was 65. I worked for the company 48 years. I enjoyed it. Particularly, I enjoyed the people I worked with. Some of them I remember as hard people to work with but damn good people to be with. One boss, I remember, was a terror on the job. If you made a mistake (and we did,) you got hell for it.

I never drank a drop in my life. I avoided drink all my life, although I may have a glass of wine with dinner.

I was born in 1894 in Springhill. My father was a coal miner in the Springhill and in the Joggins. It's so long ago that I started in the mines that I can't remember too well. I was very young when I quit school – about 12 years old. I went in the pit at the Hub and worked there for about a year, then I went back to school.

After I worked in No. 2 Mine, I went to Chicago and enrolled in electrical school. I was there six months and I got a terrible thirst. I went to the principal and told him we had sickness at home in Cape Breton and I would have to leave. Could I get part of my tuition to take me to Cape Breton? He gave me all my tuition so I spent that and then I got a job washing dishes in Chicago. I came home and started in the pit again.

At the time, the thing was to go to Boston. I was working in a garage in Boston when I met my wife at Scotch Corner (that's where all we Cape Bretoners would congregate.) We lived about 10 years in Boston. We got married and two of our children were born there.

During the Depression, we got itchy feet and came home. I went to work in No. 10 Colliery in Reserve Mines. Oh yeah, I liked being a coal miner better than the jobs I had in Boston.

I was working in No. 20 Colliery when I hurt my back – so I quit the pit. After I got feeling better, I was standing down on the wharf. An old man from Bridgeport, he was in a small boat and he asked me, "Do you want to go out fishing?" It was blowing a nice little breeze so I said, "Sure." And we got a nice catch of fish. I've been a fisherman since, although I stopped fishing in 1975. I liked fishing better than coal mining because it's in your blood. Being a fisherman is just like being an alcoholic.

I used to run away from school and come down here to go out with the old fellows on small sloops with just a jib and a mainsail. Sometimes we'd have bad weather. Sometimes we'd get out there and the wind would die and you'd be there for hours.

I used to go out fishing alone. I brought swordfish and 250-pound halibut in alone. I brought in one of the biggest tuna fish that was ever brought into Glace Bay Harbour. At one time, there were so many boats in Glace Bay Harbour that you could walk from one side to the other, just on the boats.

I thought that when I quit the coal mine, we'd starve to death. But I've done just as well.

BILLY CORKUM

RODDY MacDONALD

I retired in 1969 — my last day was the 22nd of December. They were phasing out the mine but wanted me to stay on. I said, No, I got a ticket bought for Toronto and I'm not cancelling it. I said, I've had 50 years and that's it.

I started in 1920 in No. 4 Caledonia. It was right across from here. It was no trouble to get work then — all you had to do was go over and see the manager.

My first job was trapping. The next job for a young boy like me was driving, hauling the coal out to the landing. When I first started driving, I was a little guy about this high and I had a great big fat horse, strong-hauling out of the deeps. I can remember that horse, by God, like it was yesterday.

We used to get paid based on the numbers of boxes we hauled: the more coal you hauled, the more money you made. The check weighman would count the boxes. You had to be ambitious. It was a system designed to make everyone work harder.

Most of my working life was in Caledonia. I became the underground manager at Caledonia. The decision to close that mine was made very hastily. It was a beautiful mine and should never have been closed. After Caledonia closed, I went to No. 26 Colliery as underground manager. I worked there five years, then quit. I didn't have to quit, they wanted me to stay.

The most money I got as manager was thirteen thousand dollars a year. Today, a manager gets so much money, he doesn't know what to do with it.

I had three trips out west. I was only 17 the first time I went. The young fellows from the mine all decided to go. The manager would holler, I'll never hire you again. But you always got on again. A lot of boys went and stayed out. Those were rough times. God, it was rough on those trains. Fights! There wasn't a window left in the train by the time you got to Winnipeg.

HERB PIKE

I was born in 1910 in Glace Bay. I started in the Florence Mine with my father, driving the main deeps with a punching machine. That was the first machine driven by air. Three dollars a day was what you got because you were called a 'helper.' When I first started working with my father, I wasn't quite 16.

I worked 43 years in the coal mines. My father worked there for 52 years. I was on contract all my life. I never carried a lunch pail because I never had time to eat. I don't want to brag but I was approached in 1969: they said, "Would you come back? We'll give you a good job looking after the arches in No. 26 Colliery." I said, "No way."

When the Florence Mine closed, I was transferred to the Princess Colliery. Then I went overseas and, after the war, I went to the gold mine in Kirkland Lake. I didn't stay long there.

Then I started in No. 20 Colliery. I did all the stone work along the rake roads. I was an arch boomer. I took a lot of pride in my work in the mine.

Why did I stay in coal mining? I'll have to tell the honest truth: I stayed in coal mining because I liked it. I liked it. Most of the people who'd answer that question would tell you I stayed because I had no alternative. But when I took jobs in other areas, there was something that always bothered me – and that was the lack of companionship. We used to have singsongs and all that stuff, five or six fellows together, and it was a lot of fun. You worked hard but there was some fun and the days went by. In fact, you looked forward to it. I was president of the Glace Bay Pensioners' Club for four terms.

PETER BYZAN, RETIRED COAL MINER & HERB ZORYCHETA, RETIRED MINING ENGINEER

Oh, I was better looking one time than now. I had blond, curly hair. The little bit that's left has now turned white. I started mining in 1915 and I was pensioned off in 1961 – that's a total of 46 years in the mines. I came from the old country. When I was born, there was no Poland; the part I came from was Austria. My father was Ukrainian and my mother was Polish.

I started in No. 2 Colliery. I'd never seen a mine before or been underground. I worked a short time in No. 2 Colliery, then I went to Caledonia No. 4. I stayed in Caledonia until I retired. They wanted me to go to No. 26 Colliery, but a friend of mine said, Don't. Day after day, I got used to mining. I worked with a Polish fellow. We worked together 27 years without a fight, always in room-and-pillar mining.

Today, miners don't work like they used to. I worked with the hand shovel for 40 years, then I did some steel booming. There are no pan shovels in the mine now, where before there was nothing but pan shovels. – *Peter Byzan*

I was born in Glace Bay in 1922. I served in the RCAF and graduated in science from Saint Xavier University. Then I joined the Department of Mines in 1951 and remained there until 1972. I had extensive experience in mining in various parts of the country; much of my work was associated with coal mining in Pictou County and Cape Breton.

For most of my time with the department, I was the federal government resident mining engineer in the Sydney area. Since 1972, I've been director of Mining, Engineering and Inspection for the Nova Scotia Department of Mines. Caledonia was the first mine I entered as a young kid – it was one of the cleanest, tidiest mines you could see. And there was a comradeship in the Caledonia Mine that I've never experienced in any other mine. My father was driving a horse underground in Caledonia and there was a roof fall. My father jumped into the box – the box saved his life – the horse was killed. I married a girl from Springhill. We have one daughter and three sons. – *Herb Zorycheta*

JOE ZORYCHETA

bought a 1959 Cadillac and I drove it right up until last year. Now I'm walking downtown to do shopping. I grow grapes, apples, plums and pears in this yard. Have you ever seen grapes growing in Cape Breton?

I was born in Poland in 1903. I got into the Caledonia Colliery when I was 15 – I told the manager who hired me that I was 16. I started trapping for 70 cents a day. I worked 22 years at Caledonia Colliery and I done every job in the mine. I worked in No. 12 before the explosion. I think that explosion killed about 70 men. I left on a Saturday and, four days later, they had an explosion.

During the Depression years, the mines were operating only one or two days a week and the men couldn't afford to buy coal. So the miners started their own bootleg coal mine. It was a real operating mine and my job was to deliver the coal. The miners went down there and gave away the coal to those who couldn't afford to pay and they bartered with those who could afford to pay. For my part, I got paid with coal. If it was a short haul, out of three loads, I would get one. If it was a long haul, then out of two loads, I would get one. I ended up with a lot of coal stockpiled in my backyard.

The boss of Dominion Coal came to me one day and said, "What would you rather do? Mine and deliver bootleg coal or work for the company?" I said, "Work for the company, if it's steady work." So I became one of the principal outside contractors hauling coal, supplies and gravel for the company.

I worked in the pits until 1942. I quit because there was no work and then I started my trucking business, full-time. I operated that business until 1971. You know, I still have 13 tons of coal in my house each year.

When I worked in the pit, of course, I didn't smoke. But when I got up, jeez, I'd sit down on the bench and smoke two cigarettes. Then I'd wash, and after that, I'd have another smoke, mark my time and then smoke another one. You damn near make up for the whole 10 hours you didn't get a smoke.

I was born in New Aberdeen in 1912. Yes, my father was a coal miner and so were my brothers — not all of them, but some. We lived in a company house. During the Depression, work was very scarce so we used to go to the mine office every day, just in case there was a chance you might get a job. I finally got hired on in No. 2 Mine, right near where we used to live.

I did various jobs in the mine. Then I became an overman in No. 26 Colliery. I was there for at least 25 years. I worked a total of 35 years for the coal company. I retired in 1972. Then I worked on a project to make the Old French Mine into a tourist attraction. When the French occupied Louisberg Fortress, they used to dig coal from a small mine on the coast and take it by boat around the Point down to Louisberg.

All in all, I liked it. I could never go downtown without some guy coming up and saying, "You owe me two hours." I always used to say, "Watch the mail for a cheque." They were good men working here and, of course, the older miners didn't really need a boss. They knew where they were going to work and that was all that was necessary.

I remember one time, Angus McGuiness — he's dead now, he was from Donkin — he was downtown on Commercial Street on a Saturday afternoon, half-drunk on a bicycle and they arrested him, you know. Well, Jesus, the word spread through the pit and the washhouse. The fellows going in on the rake started imitating the police siren. All you could hear was 'Err Err Err' and it drove poor Angus crazy. — *Duncan MacNeil*

I was born in 1916 on Wallace's Road, Glace Bay. My father went to work at 11 years of age in Newcastle, England. When he was 12, he was driving a pony and the shaft on the cart broke and went right through his leg. He often showed me that scar on his leg.

My father worked his way up and finally became manager at No. 24 Colliery, in 1951. I used to go meet him as he was coming home. I'd go and get his lunch can and I'd be his buddy. He would say, "What are you going to do, Buddy?" In 1933, my dad said, "There's a good chance for a job in the pit if you want to take it." So I started in No. 24, travelling with the deputy.

Then I loaded coal until I got my first-class papers in 1940. Then I went shot-firing and then I went deputy. When my dad retired, I took his place as manager until the colliery closed. Then I went to No. 1B Mine as maintenance overman on the wall. The last eight or nine years, I was the safety supervisor. I found that job more interesting than the other jobs I'd had.

We had a bad disaster in No. 26 Colliery — twelve fellows killed in an explosion. I was the second man into the area. It's still fresh in my mind and I don't like to talk about it.

Duncan and I were supervisors at a time when hours and time of work didn't mean a thing. Seven days a week and no extra pay for it. I became president of an association of mine officials. Our committee talked to the powers-that-be and we ended up with a 40-hour week. Any more than that, we got time-and-a-half.

Duncan and I have been buddies for over 30 years. When he retired in 1972, he said, "This will be the end of us. You'll be working and I won't. We'll just drift." But every two or three days, the telephone will ring or I'll ring. — *Al Atkinson*

DUNCAN MacNEIL, RETIRED OVERMAN & **AL** ATKINSON, RETIRED SAFETY SUPERVISOR

LIST OF PHOTOGRAPHS

Coal Miners of British Columbia and Yukon

Coal Miners of Alberta

Coal Miners of Alberta cont'd

125 Ruby Koflik, Payroll Supervisor, Whitewood Mine, Wabamun, 1987
126 Don Kingdon and Louie Carriere, Retired Mine Managers,
  Wabamun, 1987
127 Lyle Hobbs, Mining Engineer and David Denton, Shovel Operator,
  Whitewood Mine, Wabamun, 1992
128 Vern Tarnowski, Reclamation Supervisor, Highvale Coal Mine,
  Seba Beach, 1987
129 Jim Ferrier, Truck Driver, Highvale Coal Mine, Seba Beach, 1987
130 Randy Davis, Dozer Operator, Genesee Coal Mine, Genesee, 1992
131 Art McClure, Production Foreman, Ray LeGrow, Mine Superintendent and
  Denis Gaspe, Mine Manager, Genesee Coal Mine, Genesee, 1992
132 Charles Ostertag, Retired Surface Mine Operator, Evansburg, 1982
133 Wes Bremner, Shovel Operator, Coal Valley Mine, Coal Valley, 1991
134 Alex Matheson, Retired Coal Miner, Cadomin, 1991
135 Keith Boudreau, Shovel Operator and Don Wallace, Truck Driver,
  Gregg River Coal Mine, Hinton, 1987
136 Harry Dirk and Dalton Roy, Mechanics, Gregg River Coal Mine,
  Hinton, 1987
137 Greg LaVallee, Blasthole Driller, Gregg River Coal Mine, Hinton, 1987
138 Kyle McCracken, Dozer Operator and Dale Gravel, Serviceman,
  Luscar Coal Mine, Cardinal River, 1991
139 Chris Race, Heavy Duty Apprentice, Don Moorehead, Welder, Daryl Davis,
  Mechanic, Gordon Bancroft, Mechanic, and Kirk Mork, Mechanic, Luscar
  Coal Mine, Cardinal River, 1991
140 Anne Marie Toutant, Mining Engineer, Luscar Coal Mine,
  Cardinal River, 1991
141 Will Morton and Ray Martin, Drillers, Smoky River Coal Mine,
  Grande Cache, 1991
142 Dave Schwartz, Bill Lee and Adam Jeballa, Coal Miners, Smoky River Coal
  Mine, Grande Cache, 1991
143 Lloyd Layes, Bobby Stewart and Malcolm McNeil, Underground Mechanics,
  Smoky River Coal Mine, Grande Cache, 1991
144 Shannon Phillips, First Aid Attendant, Smoky River Coal Mine,
  Grande Cache, 1991
145 Bill Millward, Geologist, Smoky River Coal Mine, Grande Cache, 1991
146 Louis Douziech, Mine Operator, Thorhild Coal Mine, Thorhild, 1991

Coal Miners of Saskatchewan

147 Tom Mead, Shovel Operator, Poplar River Mine, Coronach, 1987
149 Gerry Beaubien, Oiler, Wayne Kelley, Mine Manager and Wes Heatcoat,
  Dragline Operator, Poplar River Mine, Coronach, 1987
150 Don Wilson, Electrician, Poplar River Mine, Coronach, 1987
151 Emerson Jones, Shovel Operator, Poplar River Mine, Coronach, 1987
152 Harvey Boles, Retired Mine Manager, Estevan, 1987
153 Johnny Wetsch, Retired Shovel Operator and Casper Fiest, Retired Welder,
  Estevan, 1987
154 Joe Hirsch, Retired Tipple Foreman, Costello Mine, Estevan, 1987
155 Jack Hill, Retired Paymaster, Estevan, 1987
156 Millard Holmgren, Retired Mine Superintendent, Estevan, 1987
157 Lawrence Stephany, Pit Supervisor, Utility Mine, Estevan, 1987
158 Dragline Pit, Utility Mine, Estevan, 1987
159 John Lochbaum, Dragline Operator, M&S Coal Mine, Bienfait, 1982
160 Jim Saxon, Serviceman, Boundary Dam Mine, Estevan, 1998
161 Bryan Daae, Training Coordinator and Joanne Brokenshire, Dozer
  Operator, Boundary Dam Mine, Estevan, 1998
162 Lance Marcotte, Tipple Operator, Bienfait Mine, Bienfait, 1998
163 Kelly Avery, Crane Operator, Bienfait Mine, Bienfait, 1998
164 Mike Matijevich, Retired Coal Miner, Bienfait, 1982
165 Metro Katrusik, Retired Coal Miner, Bienfait, 1987
167 Alex Ronyk, Retired Coal Miner, Bienfait, 1982
169 Rose and Stewart Kidd, Retired Dragline Operator, Bienfait, Saskatchewan
  and Julie Auld, Retired Office Supervisor and John "Toby" Auld, Retired
  Coal Miner, East Coulee, Alberta, 1982
170 Earling Rothe, Machinist. Costello Mine, Estevan, 1987
171 Bob Toombs, Engineer and Gary Gedak, Brakeman,
  Costello Mine, Estevan, 1987

172 Lawrence Charlebois, Tipple Foreman, Costello Mine, Estevan, 1987
173 Elmer Holsteine, Shift Foreman, Costello Mine, Estevan, 1987
174 Char Briqette Plant, Manitoba and Saskatchewan Coal Mine, Bienfait, 1982
175 Les Kingdon, Retired Coal Miner, Bienfait, 1982
177 Peter Gemby, Retired Coal Miner, Bienfait, 1982
179 Robert Long, Retired Tippleman and Stan Long, Retired Tippleman,
  Roche Percee, 1987
180 Sue and Burt Stock, Retired Shovel Operator, Roche Percee, 1982
181 Sarah and Sam Lester, Retired Coal Miner, Roche Percee, 1987
182 Gladys and Joe Wrigley, Retired Boilerman, Roche Percee, 1982

Coal Miners of New Brunswick

183 Larry Comeau, Welder, NB Coal Limited, Minto, 1987
185 Davie Betts, Firman Betts, Retired Coal Miner, Debbie Thibideau
  and Glen Betts, Minto, 1980
186 Stanley Crawford, Retired Coal Miner, Newcastle Bridge, Minto, 1980
187 Bramwell Moore, Retired Coal Miner and granddaughter, Tammy
  Moore, Newcastle Bridge, Minto, 1980
188 Billy Spencer, Retired Coal Miner, Minto, 1980
189 Frank Coakley, Retired Coal Miner, Newcastle Bridge, Minto, 1980
190 Dick Quigley, Retired Coal Miner, Minto, 1980
191 Russell Ackerman & Arthur Ackerman, Retired Coal Miners,
  New Zion, Minto, 1987
192 Frieda and Clare Roberts, Retired Coal Miner, Minto, 1980
193 Jimmy Forrester, Retired Coal Miner, Newcastle Creek, Minto, 1980
194 Rose and Manuel Cantini, Retired Coal Miner, Minto, 1987
195 Jimmy Richardson, Retired Coal Miner, New Zion, Minto, 1987
196 Gaiety Theatre, Theatre Street, Minto, 1987
197 Ed Ryan, Retired Coal Miner, Minto, 1987
198 Madeline Fila and Garnet Richardson, Retired Coal Miner, Minto, 1980
199 Fred Legere, Retired Coal Miner, Minto, 1980
200 Dominic DiCarlo, Retired Coal Miner and Shop Owner, Minto, 1980
201 Eli Goguen, Retired Coal Miner, Chipman, 1980
202 Queeni and Bill Busch, Retired Coal Miner, Chipman, 1980
203 Rubin Reed, Retired Coal Miner, Newcastle, Minto, 1987
204 Wendall Welton, Retired Coal Miner, Minto, 1980
205 Louie Madore, Retired Coal Miner, Minto, 1980
206 Harold Brown Jr, Alvin Brown & Harold Brown, Retired Coal Miner,
  Minto, 1980
207 Wellesley Hoyt, Retired Mine Operator, Newcastle, Minto, 1980
208 Surface Coal Mine, NB Coal, Minto, 1980
209 Harry Thompson, Mine Comptroller, NB Coal, Grand Lake, 1980
210 Aubrey Elliott, Maintenance Foreman and Art Thompson, Welder,
  NB Coal, Minto, 1987
211 Billie Forrester, Welder, NB Coal, Minto, 1987
212 Dwaine Barton, Welder, NB Coal, Minto, 1987
213 Eric Barnett, Hazen Brown, Walter Fulton, Cameron Knox, Eli Richards,
  Elvin Wood, Welders, NB Coal, Minto, 1987
214 Vaughn Legassie and Harold Campbell, Loader Operators, NB Coal, Minto, 1987
215 Alton Hoyt, Mine Operator, Lake Industries, Minto, 1980
216 Clarence Crawford and Roy Sonyier, Surface Miners, Lake Industries, Minto, 1987
217 Roy Mills, Retired Mine Operator, Lake Industries, Minto, 1987
218 Raymond Garon, Surface Miner, Lake Industries, Minto, 1987

Coal Miners of Nova Scotia

219 Blain Dugas, Kevin Marsch and Cliff Ligatto, Coal Miners,
  Prince Colliery, Point Aconi, 1986
221 Bob Hachey, Retired Coal Miner, River Hebert , 1981
222 Romney McAloney, Retired Underground Manager, Strathcona, 1981
223 Conrad Roberts, Underground Manager, Joggins, 1981
224 Jude Melanson, Retired Coal Miner and Walter LeBlanc,
  Retired Coal Miner, River Hebert, 1981
225 Ronald Beaton, Retired Mine Operator, Springhill, 1981
227 Joe E. Tabor, Retired Coal Miner and John Laurie, Retired Coal Miner,
  Springhill, 1981

Front Cover: Coal Miners' Picnic, East Coulee, Alberta, 1991
Frontispiece: John Fry, Coal Miner, Blairmore, Alberta, 1985

ACKNOWLEDGEMENTS

First of all, special thanks must go to all the coal miners who warmly invited me to hear their stories about their lives in mining then helped me find the appropriate background for their photographic portraits.

The most difficult part of this project was making a final selection of photographs for the book. I sincerely wish there were more pages to include a photograph of every mining person I had the good fortune to meet.

Financial assistance to carry selected photography was received from Canada Council, Alberta Culture, Alberta Chamber of Resources and Manalta Coal – this help is gratefully acknowledged. All participating coal mining companies facilitated my visits to their mine sites.

Lastly, I greatly appreciate the support provided by my family and friends, both inside and outside the mining industry. Particular thanks go to a coal miner's daughter and friend, MaryAnn.

BIOGRAPHICAL NOTES

Lawrence Chrismas is the wearer of three hats: documentary and commercial photographer, executive producer of a music CD, and a coal marketer in the mining industry. He jokingly suggests that he had to maintain a day job in the industry to finance his passion for documentary photography.

Chrismas has compiled a veritable photographic history of coal mining through his portraits of retired Canadian coal miners; it is acknowledged to be the most comprehensive collection of its kind in the country. And although he admits that while he finds the mines, equipment, technology and unions of the industry to be of interest in his work, his particular affinity is for the coal miners themselves.

In 1997, Lawrence Chrismas was awarded a fellowship in the Canadian Institute of Mining and Metallurgy; he has been an active member of the CIM since 1966. He is past chairman of the Coal Division and is currently on the executive of the Calgary branch of the Institute. He served four years as the chairman of the History and Heritage Committee of the Coal Association of Canada. Currently, Chrismas is a director of the Historic Atlas Coal Mine Museum in East Coulee. Born in Alberta and educated at the Universities of Alaska and Alberta in geology and mining, he has worked 25 years in the mining industry.

Lawrence Chrismas has been documenting individuals in personal or portraiture photography since the early 1970s. He has exhibited his coal miner portraits widely in Canada since 1982 and his photographs reside in various private, public and corporate collections. His books include *Alberta Miners – A Tribute* (1993) and the 1989 *Manalta Miners* and *Minto Miners*.

NOTE

Provincial laws in Canada require that persons working in mining operations abide by the safety regulations. In most situations, workers must wear hard hats, protective eye wear and steel-capped boots. If this protective gear is absent from selected photographs taken at operating mines, the photographer accepts the responsibility for requesting the removal of hats and glasses.

COLOPHON

DESIGN: Gail Pocock, Bulldog Communications

EDITING: Kerry McArthur, Westword Communications
Maureen Geddes, Red Pencil Copyediting Ltd.
Douglas MacFarlane, Retired Coal Geologist

PRINTING: Friesen Printers

DUOTONE
SEPARATIONS: TC4 Graphics Ltd.